Minute Man National Historical Park:
Rte 2A Traffic Analysis and Its Impact on the Park's Visitor Experience

Authors

Jeff Bryan, Volpe Center
David Spiller, Volpe Center
Scott Peterson, EG&G
Francis Ford, EG&G

Date

June 28, 2002

Prepared by

U.S. Department of Transportation
Research and Special Programs Administration
John A. Volpe National Transportation Systems Center

Prepared for

U.S. Department of the Interior
National Park Service Minute
Man National Historical Park

Volpe National Transportation Center

NOTICE

Neither the United States Government nor any agency thereof, nor any of their employees, makes any warranty, express or implied, or assumes any legal liability or responsibility for the accuracy, completeness, or use of any information, apparatus, product, or process disclosed. Reference herein to any specific commercial product, process, or service by trade name, trademark, manufacturer, or otherwise, does not necessarily constitute or imply its endorsement, recommendation, or favoring by the United States Government or any agency thereof. The views and opinions of authors expressed herein do not necessarily state or reflect those of the United State Government or any agency thereof.

Volpe National Transportation Center

Acknowledgements

The authors wish to thank the numerous organizations and individuals who graciously provided their time, knowledge and guidance in the development of this report. Those of particular note are listed below.

Stakeholders
- Nancy Nelson, NPS
- Gregory Cravedi, Hanscom AFB
- Efi Pagisas, CTPS
- Mary McShane, CTPS
- MHD Record Division
- Town of Concord
- Town of Bedford
- Town of Lexington
- Town Lincoln
- Steve Lawson, UVT
- Robert Macintosh, NPS
- Tom Casey, NPS
- Bill Brooks, NPS

Volpe National Transportation Center

Table of Contents

1.0 Purpose of the Study 1
   1.1 Key Concepts 1
   1.2 Method of Analysis 2
2.0 Historical Perspective 2
   2.1 Battle Road (Historical Context) 2
   2.2 Battle Road (Geographical Context) 4
   2.3 Minute Man National Historical Park (Historical Context) 4
   2.4 Minute Man National Historical Park (Geographical Context) 5
   2.5 Route 2A 6
3.0 Visitor Experience 6
4.0 Traffic Conditions 7
   4.1 Roadways Impacting the Park 9
      4.1.1 Rte 2A (Concord / Lincoln / Lexington) 9
      4.1.2 Hanscom Drive (Lincoln) 10
      4.1.3 Lexington Road (Concord / Lincoln) 10
      4.1.4 Bedford Road (Lincoln) 10
      4.1.5 Airport Road (Lincoln / Lexington) 10
      4.1.6 Massachusetts Avenue (Lexington) 11
      4.1.7 Old Mass Avenue (Lexington) 11
      4.1.8 Route 128/Interstate 95 (Lexington) 11
      4.1.9 Virginia Road (Concord / Lincoln) 11
   4.2 Rte 2A Roadway Design 12
      4.2.1 Functional Classification 12
      4.2.2 Functional Speed 16
      4.2.3 Traffic Volumes 16
      4.2.4 Design 16
   4.3 Rte 2A Traffic Volumes: Historical, Current, and Projected 18
      4.3.1 Historical 18
      4.3.2 Current 20
      4.3.3 Projections 23
   4.4 Trip Patterns: Origins, Destinations, & Flows 23
      4.4.1 Origins 23
      4.4.2 Destination 24
      4.4.3 Flows 24
   4.5 Analysis of Rte 2A Traffic 25
      4.5.1 Volume to Capacity 33
      4.5.2 Intersection (Level of Service) 33
      4.5.4 Pedestrian Access 34
      4.5.4 Speed Issues 35
      4.5.5 Accident & Safety 36
      4.5.6 Noise Levels 37
   4.6 Current and Future Considerations 39
      4.6.1 Planning Process through MAPC 40
      4.6.2 Planning Process through HATS 40

Volpe National Transportation Center

  4.6.3 Future Transportation Projects 43
  4.6.4 Public Transportation 45
5.0 Optimal Traffic Volume 47
6.0 Alternatives 49
 6.1 Rte 2A Redesign Options: Positive Impact on the Visitor Experience 49
 6.2 Designation Status Options as a Means to Improve Rte 2A 55
 6.3 Rte 2A Redesign Options With a Possible Negative Impact on the Visitor Experience 58
 6.4 Framework for a Traffic Management and Road Re-design Plan 60

1.0 Purpose of the Study

Minute Man National Historical Park is located in the towns of Concord, Lexington, and Lincoln Massachusetts. It spans over 900 acres of mixed land use that includes numerous historic sites, structures, trails, and landscapes associated with the American Revolution. Over 1,000,000 people visit the park each year to experience, understand, enjoy, and appreciate the features contained within it. One of the newest attractions that the park has made available to the public is the Battle Road trail. It consists of a five- mile interpretive trail that follows portions of the historic Battle Road where the war for Independence began and allows the visitor to view the other landmarks in the park in the same way as the Minuteman of yesteryear did. The primary means to access the park and Battle Road is Rte 2A which run in the same east-west alignment as the Battle Road Trail. In addition to providing access to the park it provides access to Hanscom Air Force Base, Hanscom Regional Airport, a school, local residents, local businesses, and lastly as a cut through for trips originating west of the park to points east within the Metro Boston area. The problem that this report tackles is determining how to best utilize Rte 2A and the surrounding roadways to access Minute Man National Historical Park and Battle Road while minimizing the impact of other trips on the visitor experience. This study tries to answer this question in two ways.

- Identify a maximum daily traffic level on Rte 2A that could provide the best possible traffic movement that benefits the visitor to the park while still allowing the other trips to use this roadway.
- Propose several options that can be combined or done separately to either help maintain the desired traffic level and minimize the traffic impact on the visitor experience.

1.1 Key Concepts

The visitor experience and their satisfaction were defined as a goal in the Minute Man National Historical Park Annual Performance Plan, dated 2002. Mission Goal 2A describes it as allowing the public the opportunity to enjoy the park and provide for a memorable visitor experience. Two key points support this goal.

- The visitors should safely enjoy and be satisfied with the availability, accessibility, diversity, and quality of the parks facilities, services, and the appropriate recreational opportunities at Minute Man National Historical Park.
- The visitors and the general public should be able to understand and appreciate the preservation of Minute Man National Historical Park and its resources for future generations.

The traffic on Rte 2A and the surrounding roadways can impact the visitor experience in several ways.

- Traffic Congestion: volumes, delays, and vehicle speeds can be measures of congestions impact on the visitor experience
- Travel Patterns: This can be measured by examining the origins and destinations of the vehicles that use the roadways in the study area and then understand why they choose the route they did using traffic flows.
- Noise: Noise is generated by traffic, airplanes, and by the people adjacent to the park property and can be measured by its loudness.

- Safety Concerns: Traffic on the roadway and the design of the roadways themselves can contribute to safety concerns for the park visitor and be measures by accident reports
- Environmental Concerns: Several issues such as air quality and drainage from the roadway can detract from the visitor experience.
- Visual pollution: Personal feelings about viewing automobile traffic adjacent to historic sites that predate the use of the automobile.

1.2 Method of Analysis

This analysis will explore the linkages between Minute Man National Historical Park, Battle Road, Rte 2A, and the surrounding roadways by identifying the historical context that created them, exploring the spatial relationship between them, then quantifying the concerns relating to traffic congestion, travel patterns, noise levels, and safety. All the while trying to understand the impact they would have on the visitor experience. Even though environmental concerns and the issue of visual pollution exist, this study won't tackle them due to a lack of data, time, and financial constraints that exist on this project.

The results of the analysis on traffic congestion, travel patterns, noise, and safety will be used to develop a traffic volume threshold answers the first question. Once the problem areas of the roadway have been identified, several possible solutions will be presented to address the second question based on the current state of the practice in transportation planning and traffic engineering.

2.0 Historical Perspective

In order to understand the relationship of the Battle Road, Minuteman Park, and Rte 2A individually and together; one first must understand the historical and geographical relationship between them. There are several key questions that should be answered using these perspectives, why is Battle Road important, what is the relationship between the park and Rte 2A, and lastly how does the alignment of Rte 2A and Battle Road impact one another?

2.1 Battle Road (Historical Context)

"Battle Road" is the designation used on a portion of the route from Boston to Concord taken by British soldiers on April 19, 1775. The entire route passes from Boston through Cambridge, Arlington (formerly Menotomy, West Cambridge), Lexington, and Lincoln to Concord.

The British were marching to Concord to seize arms and ammunition stockpiled by the colonists. Tension had been building between England and the colonies since the end of the French and Indian War in 1763. New taxes and punitive acts on the colonies by Britain and alternating boycotts and protests culminated in the Boston Tea Party. King George III responded to the Tea Party with a series of punitive acts known as the "intolerable acts".

Britain had been aware for some time that the colonists were stockpiling arms. They successfully seized stored arms in Cambridge, but a subsequent seizure attempt at Salem was unsuccessful. Britain wanted to squelch unrest in the colonies without war. General Thomas Gage, governor of the Massachusetts Bay Colony, knew that any further attempts at seizing arm would require secrecy.

On the evening of April 18, 1777 650 to 900 British troops, under the command of Lt. Col. Francis Smith with orders from General Thomas Gage, assembled on Boston Common prepared to march to Concord. Word of this leaked to the colonists. As soon as the British began to leave Boston, Paul Revere and William Dawes and other alarm riders road ahead of the British and warned the colonists. Word spread rapidly. At sunrise, as an advanced guard of about 200 British marched up Massachusetts Avenue, 77 militia men were already assembled on the Lexington Green waiting for them. No one knows who fired the first shot. The British, in response but without clear orders, fired at the militia, killing eight and wounding 10 colonists, taking no casualties themselves. The militia dispersed ending the first of many engagements that day.

The British continued to Concord along Massachusetts Avenue. By this time 150 colonists had assembled in Concord. As the British moved towards the North Bridge, the colonists crossed ahead of them and took positions above the bridge. This left the British free to search the town center. About 90 British solders secured the bridge while others went to search Barrett's Farm. British in the center of town began burning and destroying found supplies. The militia, whose ranks grew to over 400, thought the town was burning and advanced back towards the bridge.

The British retreated across the bridge and started to remove the bridge planking to keep the militia at bay. The British fired into the river, then at the militia, killing two and wounding several others. Maj. John Bartlett then ordered the militia to return fire, killing two British soldiers died and wounding several others whom later died. The British retreated to the Concord Center. The British regrouped and began their return march to Boston by noon.

The colonists, now 1,000 strong gathered near the house of Nathan Meriam (Meriam's Corner). Here the British gathered to cross a small bridge in Bay Road (Now route 2A). The colonists opened fire, beginning the running battle to Lexington Green.

The militia followed the retreating British, firing at them at will in a running battle from Meriam's Corner to Lexington green. At Lexington Green Lord Percy, 1000 troops and two cannons dispersed the minutemen and allowed the British to reorganize and continue their retreat to Boston. Although the retreat from Lexington to Boston under Lord Percy was more orderly, the minutemen continued to follow the British shooting at them from the wayside. The retreat was a success but at great cost to the British. At day's end 273 out of 1,700 British soldiers were either killed, wounded or missing; 93 out of approximately 3,700 colonists died. The colonists closed all land approaches to Boston. This was the first encounter where the provincial militia inflicted casualties on the British troops. The American Revolution for Independence had begun.

The events that took place April 19, 1775 on "Battle Road" are not only important to the history of the United States, but of the world. At that time Britain enjoyed a powerful army and the world's greatest navy. American Revolution led to the creation of the United States of America and its Constitution. A constitution that defined the principles such as the right to vote, the right of free speech, the right of free assembly, life, liberty and the pursuit of happiness. The principles have been a model for new democracies throughout the world for over 200 years, hence the significance of that day on world history.

The battle started on Lexington Green and the North Bridge at Concord and ended in Boston. Almost the entire route of the British retreat from Concord to Boston is heavily developed and therefore quite different from 1775. The section in Lincoln and west Lexington, the stretch where most of the major events took place, has seen the least development and so presents the best opportunity to recreate an eighteenth century environment with the least amount of work.

2.2 Battle Road (Geographical Context)

The Battle Road Trail is a 5.5 mile interpretive, multi-use trail that provides safe pedestrian, stroller, wheelchair, and bicycle access to Trail goers. Visitors have the opportunity to enjoy vistas of historic farming fields, protected wetlands, forested areas as well as where these natural resources have interfaced with human society; stone walls, historic structures, homes, foundations and more!

Consisting of a stone-dust surface, the trail contains several sections of the original Battle Road that have been restored to the 25-foot wide portions of the "historic" Battle Road used on April 19, 1775. These have been linked together by 7-foot wide sections of dirt and stone dust paths that try to traverse many of the same landscape features that have been identified with the Revolutionary war period.

Winding through the park from Meriam's Corner in Concord to Fiske Hill in Lexington, visitors are guided by a series of outdoor exhibit panels and granite mile markers, historic homes, stone walls, remnants of structure foundations as well as some of the park's natural features. The trail, through the help of park rangers, helps the visitor interpret both the "natural" and "human" story of the area.

2.3 Minute Man National Historical Park (Historical Context)

Many of the sites where the first fighting of the Revolutionary War took place are being preserved by The National Park Service or by the local communities. It took a long time for people to agree on what to preserve how they should commemorate the events of the first battle. The first monument erected at the site of the North Bridge was a granite obelisk, dedicated on July 4, 1837. The monument was built on what most people considered the "British" side of the bridge because there was no longer a bridge at the site. It had been torn down years before. By the time Americans were ready to celebrate the *centennial* of the start of American Independence, the people of Concord had solved this problem by rebuilding the North Bridge. Sculptor Daniel Chester French designed a statue of a Minute Man that was placed on the west side of the reconstructed bridge. On April 19, 1875, President Ulysses S. Grant and about 10,000 other people attended the dedication of the statue and the bridge.

As time went by Americans became more and more interested in their country's beginnings. The Minute Man, as depicted by Daniel Chester French, became a symbol of America's readiness to fight to preserve its liberties. During World War II, people all over the United States became familiar with the statue when it was used on posters to encourage Americans to support the war effort Long before the *bicentennial* celebration in 1975, committees were formed to plan for the crowds of visitors. These committees decided that a national park should be formed to preserve the places where Americans first fought for their liberty.

The United States Congress and President Dwight D. Eisenhower created Minute Man National

Historical Park on September 21, 1959. Today, Minute Man National Historical Park is one of more than 360 places managed by the U.S. National Park Service. It is the job of the National Park Service to preserve and protect these places and to make sure visitors understand why the places are important to all people.

Minute Man National Historical Park was established in 1959 but it wasn't until 1996 any comprehensive development of support facilities, and historic preservation took place. This was due to the large number of land parcels that were still privately owned. Once these land parcels became available, the park created a visitor center, parking areas, and began preservation efforts to maintain or restore the historic structures in the park. The majority of these sites remained inaccessible to the general public until the Battle Road Trail was created as a means to link these sites together following the original Battle Road where possible. The major problem with following original Battle Road was the current location of Rte 2A. The traffic on Rte 2A posed a safety concern so the trail was designed to be several yards away and parallel Rte 2A on its northerly side, only intersecting Rte 2A to connect with parking sites, to cross intersecting roads, and link adjacent historical sites. The park is currently in a period of planning for this development and preservation work by producing the Annual Performance Plan and updating its Strategic Plan. Minuteman Nation Park will be identified as the Park in the rest of this report.

2.4 Minute Man National Historical Park (Geographical Context)

The Park contains 967 acres of land that straddles Rte 2A and Lexington Road through the towns of Concord, Lexington, and Lincoln. Rte 2A is aligned on an east to west direction and is the primary means of access to the parks historical sites, visitor center, and parking facilities. Several north-south orientated streets serving local residents, businesses, and the airports that exist on the parks northerly side intersect the road a regular intervals. Most notably from east to west are Rte 128 / I-95, Airport Road, Hanscom Drive, and Bedford Road. Another major road running parallel to and south of Rte 2A is Rte 2.

The trail can be accessed from a number of different parking areas located off Route 2A including the Minute Man Visitor Center. The trail helps links numerous historic sites and landmarks, several of which are presented below to aid the reader.

Historic Sites and Facilities in the Park
1. Meriam's Corner
2. North Bridge, Concord
3. Fiske Hill, Lexington
4. Hartwell House
5. The Wayside Inn
6. Battle Road
7. Historic Farming Fields
8. Captain William Smith House
9. Ebenezer Fiske House Site
10. Park Visitor Center & Parking Facility

There are six newly created parking areas that visitors can use to access sites in the Park and the Battle

Road Trail. They are located at the following locations from west to east.
- Orchard House, across the street from the Wayside Inn on Lexington Road
- Near Meriam's House on Lexington Road
- Near Samuel Brooks Tavern on Rte 2A
- Near the Samuel Hartwell House on Rte 2A
- Near the site of that Paul revere was capture by the British
- The main parking facility near the Visitor Center of Rte 2A

All of the landmarks and parking facilities require the park visitor to travel on some portion of Rte 2A or Lexington Road to either access them, making Lexington Road and Rte 2A integral components of the visitor experience.

2.5 Route 2A

This roadway provides access to the park and provides mobility to park visitors, local residents, businesses, and shoppers through the study area. A good portion of the routes alignment dates back to the Revolutionary war and currently shares the same Right-of-way that the original Battle Road had in several locations. This roadway has one lane in each direction and allows for speeds of up to 40 mph while serving as an access to the park, surrounding homes, and businesses.

It is designated by the state as an alternate route to Rte 2 that lies to the south and is connected at Crosby's Corner by a small section of road called the Rte 2A Bypass. This connection allows more people to use Rte 2A for access but also increases its use as a cut through street. Historically the Rte 2A designation didn't traverse the Bypass as you go westward from Boston, it continued on Lexington Road through Concord Center. Given the congestion problems in downtown Concord, this path for Rte 2A was changed to avoid downtown Concord and connect with Rte 2 at Crosby's Corner. Given the importance of Rte 2A to the park and others, this traffic analysis will focus on identifying the problems and offer options to improve the roadway from the Park's perspective.

3.0 Visitor Experience

The park, its features, and the landscape help shape the visitor experience. Minute Man National Historical Park identified five external factors that could impact the visitor experience in their Annual Performance Plan.

- Increased traffic in the park
- Increased flights at the airport
- Construction Impacts
- Community emphasis on conservation and preservation
- Trends affecting operation

The park has identified the factors that will impact them but not the degree to which they will impact the visitor experience. Transportation professionals have standard measures for analyzing traffic like the Level-of-Service (LOS), accident analysis, and average speeds but these measures don't necessarily reflect the perceptions of visitors to a national park.

Several universities have undertaken research to understand how traffic and crowding affect the visitor experience and use this to determine what is acceptable and what isn't and how it varies by park type. Most of the research that has been done has focused on large parks where the landscape and scenery are the focus, not historical sites. The University of Vermont has been actively involved in these studies and has examined traffic at Acadia, Yosemite, and Cape Cod National Seashore. This research used surveys that attempted to equate a Level-of-Service with the visitor experience. Given the differences in location, size, and character of these parks, it is difficult to draw conclusions from them that can be applied to Minute Man National Historical Park.

Some surveys have been done at Minuteman but these were for economic research not for traffic analysis. Some of the results of these surveys indicated the following:

- During July and August 60% of visitors were in the Lexington Concord area for the first time. Only 37% were first time visitors during December
- Over 75% of visitors to Minuteman were part of a day trip
- Most multi-day trips to the Lexington-Concord area were 2.0 days during the summer and between 1.5 and 1.8 during the fall and winter months.
- Most visitors on multi day trips stayed with friends or relative. Visitors used area hotels mostly in July (25%) and lesser during other months
- The main reason for visiting the park is interest in the American Revolution followed by interest in American culture and American Literary Heritage.
- Most respondents had no trouble finding the park.
- Most visitors reside in Massachusetts.
- 80% to 90% are visitors arrive at the park by a private or rented automobile. The average party size ranges from 4.4 to 5.6 persons.
- Less than half of the visitor surveyed said they would use a "convenient, reasonably priced trolley service connecting Lexington and Concord tourist attractions". 32% said they would not. The remaining visitors were not sure.

A survey similar to the one the University of Vermont did at the other National parks done should be conducted at Minute Man National Historical Park in order to better understand how traffic impacts the visitor in an urban and historical setting.

4.0 Traffic Conditions

Traffic has become an important issue for Minute Man National Historical Park. Traffic impacts people ability to get to the Park as well their ability to enjoy it. These impacts can manifest themselves in terms of the traffic congestion, travel patterns, traffic noise, environmental concerns, visual pollution, and safety issues for everyone in the area as well as in the park itself. Traffic congestion can increases travel times, annoys the traveler, and limits the amount of free time they have. The traffic patterns are important because they affect the travel choices and routes people choose to take to get to their home, work, school, shops, and the Park. Traffic noise is a form of pollution that can affect a person's health, communication, concentration, as well as be an annoyance to the visitors of the Park. Environmental impacts generated by vehicles and roads can affect drainage, wildlife, and the air quality of an area. Large numbers of autos can create visual pollution that can negatively impact the experience of visitors to the park. The safety concerns relate to the way vehicles, pedestrians, bicyclists, and wildlife interact with the roadway based on its design, speed, and its use.

In order to determine how the traffic conditions on Route 2A impact the Park, a seven-step methodology will be used.

1. The first step consists of identifying all of the major roadways that contribute traffic to Route 2A and explain what their impact is on this roadway.
2. The second step identifies the criteria that govern its design.
3. The third step will examine the traffic volumes along Route 2A in the past, present, and then into the future.
4. The fourth step will examine what the travel patterns are on the roadways and determine how efficient or effective they are in regards to the mobility of the visitors to the Park.
5. The fifth step involves examining congestion levels by time and location.
6. The sixth step involves an examination of the noise levels that are occurring along Route 2A due to traffic and what impact this may have.
7. The seventh step looks at the safety of people traveling on Route 2A by foot, bike, and car. Once we understand how Route 2A functions, we can make an assessment of what the problems are and what traffic levels along Route 2A best accommodate the Park experience.

Several key points that this analysis discovered are:
- Average daily traffic volumes (ADT) on Rte 2A have increased by a factor of 4 in the last 40 years, going from an ADT of 5,000 in 1960 to 20,000 ADT in 2000.
- Rate of traffic growth has slowed in the last decade, possibly due to the roadway being saturated during the peak hours of congestion or changes in demographics of the area.
- No more than 1.4 percent of the two-way traffic on Rte 2A is visitation to the park, with the majority being split between local trips and trips traveling through the corridor.
- Speeds in access of the speed limit make access to and from the Park's gateways and parking facilities difficult.
- More than 50% of the eastbound AM peak hour traffic on Route 2A, West of Crosby's Corner originates on Route 2, East of Crosby's Corner.
- More than 30% of the westbound AM peak hour traffic on Route 2A, East of Hanscom Drive originates on Hanscom Drive.
- The environmental capacity of Rte 2A was calculated

4.1 Roadways Impacting the Park

The Park's principal attractions are the Visitor Center, Battle Road, and numerous historic dwellings along it. Rte 2A is the primary access to all of these and is the focus of this study. There are several roads that feed vehicles onto Route 2A that will be examined as well. The goal is to understand how they work together as a system. An examination of their design, the trip patterns on them, markets that the road serves, and levels of use will help us understand whether the use matches the need. Please refer to Figure 1 for the location of these roadways. This figure shows the six locations presented were selected based on determining what data was available from prior research and selecting points that can help the reader understand what is happening on Rte 2A. For a comparison of Average Daily Traffic (ADT) volumes on selected roadways, please refer to Figure 2. This figure compares ADT volumes on key roads in the study area in 2000. As we can see Airport Road has under 5,000 ADT, similar to Bedford Road. Lexington Road and Hanscom Drive are twice that volume with 10,000 ADT. Rte 2A at two locations are twice the volume of those roadways, with an average of 20,000 ADT. Rte 2 is more than twice that of Rte 2 with 50,000. It is important to note that all of the roadways, except Rte 2 have just one lane in each direction. Rte 2 has two lanes. When one compares the volume per lane Rte 2 and Rte 2A aren't that different.

Figure 1: Study Area and Count Locations

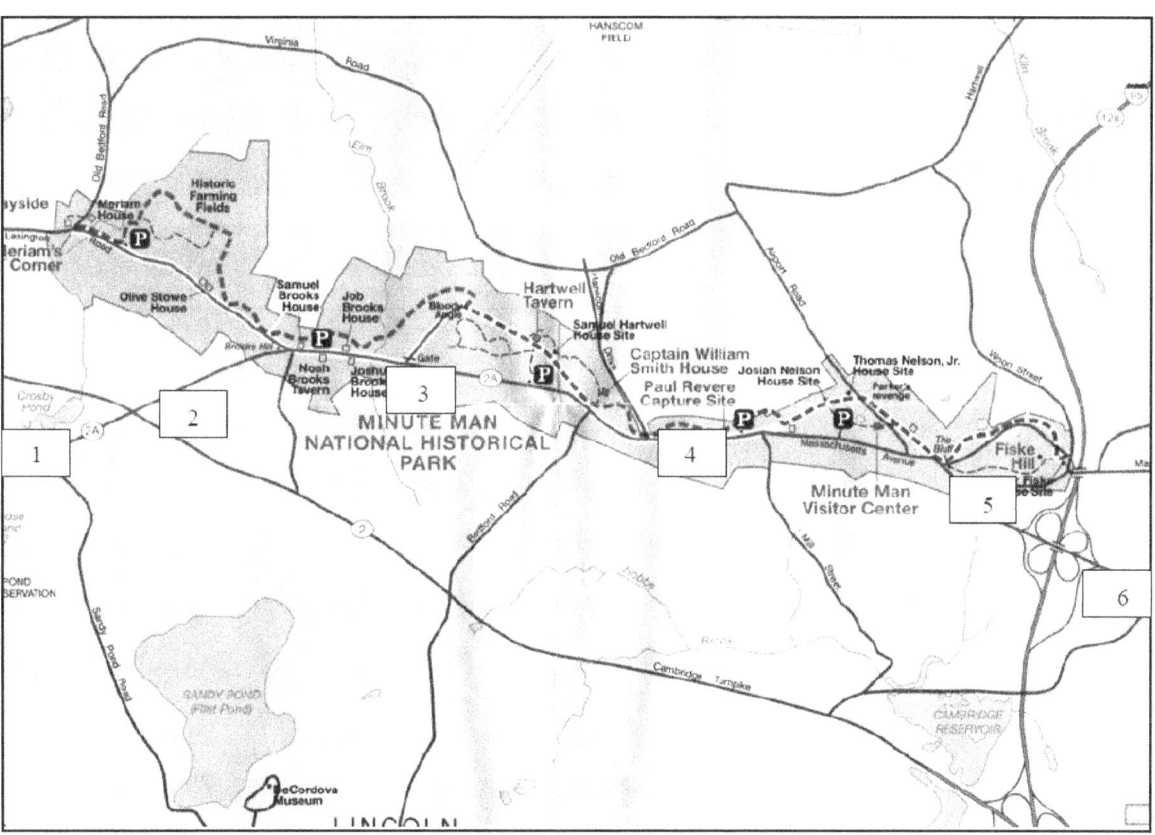

Source: National Park Service, 2001

4.1.1 Rte 2A (Concord / Lincoln / Lexington)

Route 2A provides east-west access through the Park connecting Route 2 at Crosby Corner with Route 128 and Interstate I-95 at Interchange 30. The roadway is managed by the state and is an alternate route for Route 2. The segment being examined is approximately 6.5 miles long and intersects with nine other streets along it. The towns Route 2A passes through from West to East are Concord, Lincoln, and Lexington. The Park has four parking facilities along Rte 2 and several historic sites that have their access of off Route 2A. There are a couple of flashing traffic lights, no street lighting, and several signs identifying speed limits, and Park entrances. The speed ranges from a high of 40 mph to a low of 25 at a couple of the intersections. The current function of this roadway is to provide access for local and through trips West of the Park to Interstate I-95 / Route 128, allow access to the Park, and to access the Airport. The land use along the roadway consists of scattered residential dwellings, Parkland, a school, some commercial establishments and several parking lots for visiting the Park.

4.1.2 Hanscom Drive (Lincoln)

Hanscom Drive is a north-south access road of off Route 2A to the Airport. The Airport consists of the Regional Airport as well as the Air Force Base. The road is located in Lincoln. It consists of 2 median separated lanes in each direction. The roadway is approximately 0.6 miles long and has 2 major intersections. Old Bedford Road and Vandenberg Drive are east-west roadways that intersect Hanscom Drive at its Northerly extent. Rte 2A intersects it at its southern most point. A traffic officer directs traffic at the intersection of Route 2A on weekdays between 3:00 PM and 5:30 PM. The roadway has no traffic lights, no lighting, and no signage. The average speed limit for this roadway is 30 mph. The road functions as a cut through to the Airport and the business located there. The land use along this roadway consists of parkland, several historic structures with some abutting residential dwellings.

4.1.3 Lexington Road (Concord / Lincoln)

Lexington Road is an east-west route stretching between Concord center to the west and Route 2A at Noah's Brook Tavern. It consists of 1 lane in each direction. The segment of the roadway that is being examined extends from Old Bedford Road in Concord to Route 2A in Lincoln. It is approximately 0.8 miles long. The road is located in Concord and Lincoln. The two major intersections that are on this roadway are with Old Bedford Road on the northern extent forming a T intersection and with Route 2A at its southern extent forming a Y intersection. The road functions as a local access to points east along Rte 2A or Rte 2. The roadway has flashing traffic lights, limited lighting, and minimal signage. The average speed limit is 30 mph. The land use along the roadway is primarily residential with parkland and some historic structures.

4.1.4 Bedford Road (Lincoln)

Bedford Road is a north-south way linking Route 2 to the south with Route 2A on the northerly extent. It consists of 1 lane in each direction. The segment being examined is 0.7 miles long. The roadway is located in Lincoln. There are 2 major intersections on this segment that need to be mentioned. The first is with Route 2 that is signalized. The second is with Route 2A forming a T intersection. This roadway functions as a cut through for local trips destined to or coming from the Airport and the business located there. The roadway has traffic lights, no street lighting, and limited signage. The average speed limit is 30 mph. The land use consists of scattered residential dwellings and some parkland abutting the roadway.

4.1.5 Airport Road (Lincoln / Lexington)

Airport road is a north-south route that connects Vandenburg Drive in the north with Route 2A in the south. It has 1 lane in each direction. The roadway is 0.8 miles long and is located in Lincoln. There are 2 major T intersections, one with Vandenburg Drive. The second is with Route 2A. Neither one has signals. Airport Road functions as an entry point to the Airport, businesses, and residential dwellings that are in the vicinity. The roadway has traffic lights, no lighting, and limited signage. The average speed limit is 30 mph. The land use consists of scattered residential dwellings and some parkland abutting the roadway.

4.1.6 Massachusetts Avenue (Lexington)

Massachusetts Avenue is an east-west roadway that connects the Park and communities to the west with Lexington and Route 128 / Interstate 95. The roadway has one lane in each direction and the segment that is being examined is about 0.25 miles long. It is primarily located in Lincoln and Lexington. The intersection with Route 2A is undergoing signalization and was realigned several years ago. The roadway doesn't have street lighting and but does have signage identifying the speed limit and route information. It serves as a connection between residences and businesses in the Lexington area with the residences and employment centers to the west of the Park. The average speed limit is 30 mph. The land use is primarily parkland with limited residential and commercial use along it.

4.1.7 Old Mass Avenue (Lexington)

Old Mass Avenue is a winding east-west roadway that connects the Park and communities to the west with Lexington and Route 128 / Interstate 95. This roadway was a primary connection between Lincoln and Lexington but its use has declined with the opening of Massachusetts Avenue. It has 1 lane in each direction and the segment that is considered is about 0.4 miles long. The only major intersection is with Route 2A and it is managed by a stop sign. The roadway doesn't have lighting but does have adequate signage identifying the speed limit and route information. It function is to serve as a connection between residences and businesses in the Lexington area with the residences and employment centers to the west of the Park but is use has dwindled to local residents. The average speed limit is 30 mph. The land use consists parkland and some residences.

4.1.8 Route 128/Interstate 95 (Lexington)

This is an 8 lane limited access facility used for regional circumferential access around Boston. This roadway provides access to Route 2A at Interchange 30 in Lexington and Route 4/225 in Lexington at Interchange 31. Both interchanges are un-signalized and are designed in a cloverleaf shape. The average speed on this roadway is 65 mph, with ramp speeds of 30 mph. The Airport is located a short distance from this roadway between the two interchanges. The roadway has no lights, limited lighting, and significant signage. The land uses are varied, consisting of residential, commercial, Parkland, and open space.

4.1.9 Virginia Road (Concord / Lincoln)

This is a narrow winding road that connects Old Bedford in Concord with Old Bedford Road in Lincoln. This roadway has two intersections at its end points. It is 1 lane in each direction with an average speed of 30 mph. This road functions as a cut-through for people west of Concord to employment locations in Bedford and Lexington. The roadway has no lights, limited lighting, and signage. The land uses are varied, consisting of residential and commercial sites.

4.2 Rte 2A Roadway Design

The design of a highway is governed by three factors, functional classification, design speed, and the traffic volumes that the roadway can handle versus what it is actually or predicted to handle in some future scenario. These guidelines come from the American Association of State Highway Officials (AASHTO). Due to the nature of these things in controlling how a roadway is laid out, they are called "design controls. Any proposed alterations to any of these criteria on Route 2A, can effect surrounding roadways, so it is important to consider Rte 2A design in relation to the roadways surrounding it and how changing one can alter the others.

4.2.1 Functional Classification

Massachusetts classifies roads based on a system the Federal Highway Administration (FHWA) calls functional classification. This is hierarchal system of classifying roads according to the function they serve in a roadway network based on two criteria.

1. The first criterion is based on the census classification of the area, urban versus rural. Urban roadways are prevalent inside the I-495 beltway.
2. The second criterion classifies the roadway into three categories, arterials, collectors and local roads based on their function.

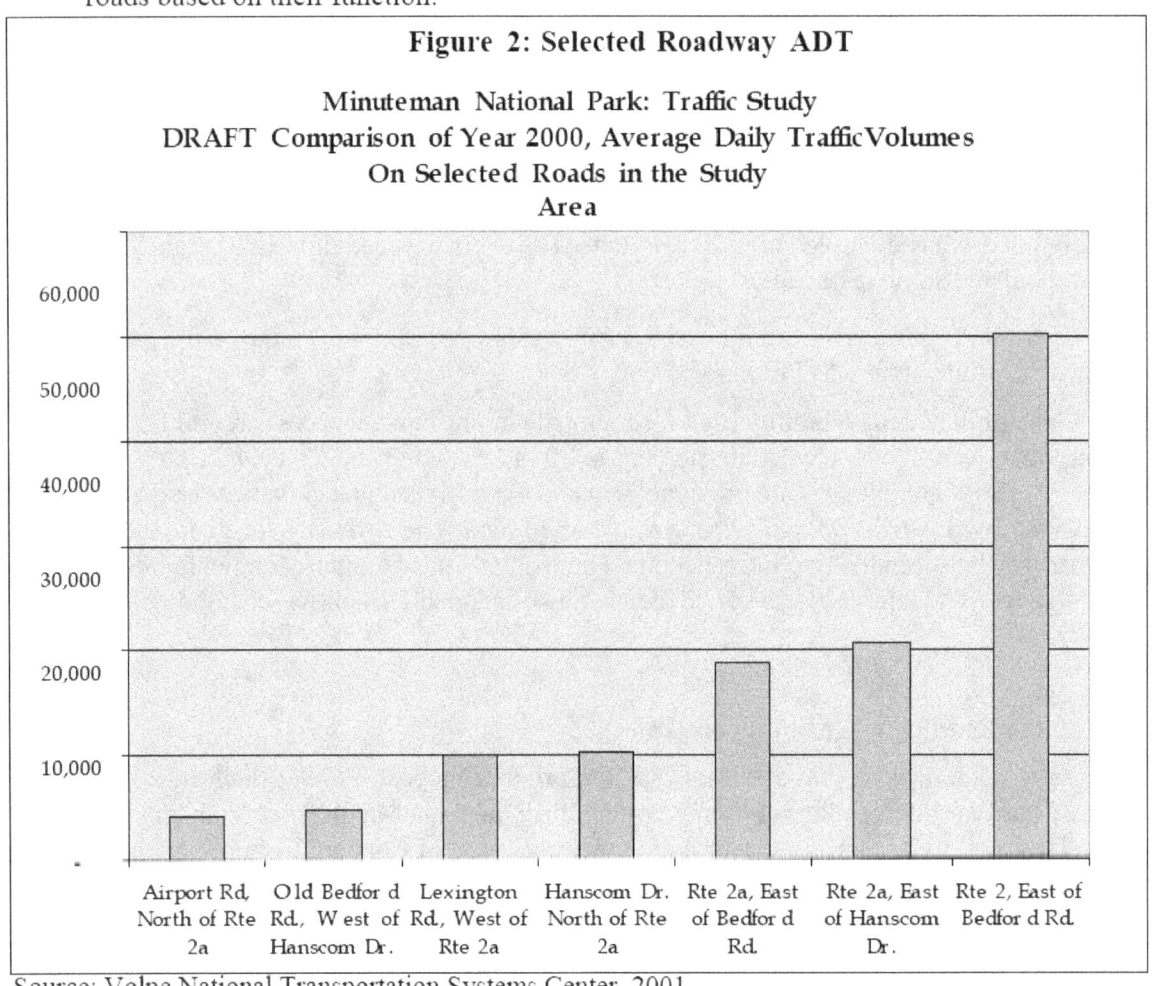

Source: Volpe National Transportation Systems Center, 2001

Function can best be defined by connectivity. Please refer to **Figure 3** for the graphical relationship between access, mobility, and functional classification. Without connectivity, neither mobility nor access can be served. Roadways that provided the greatest reach of connectivity are the highest-level facilities. Arterials can be defined by regional level connectivity. The regional movement of persons, goods and services depends on an efficient arterial system. The FHWA guidelines define arterials as "providing direct service to long trips". These routes go beyond the city limits in providing connectivity and can be defined into two groups: principal arterials (typically state routes) and arterials. Collectors both feed arterials and provide movements for shorter trip lengths. Collectors must balance high mobility and access to adjacent property. Collectors can have lower design speeds and lower levels of service than arterials.

The priority of local roads is access to abutting property so mobility and high speed is less of a concern. Each class in this system feeds traffic to a higher group. For example: local roads feed collectors, collectors feed arterials, and arterials feed interstate roads. Generally a road belonging one class terminates at a road of the same class or a higher class.

Figure 3: Mobility vs. Access in Functional Classification

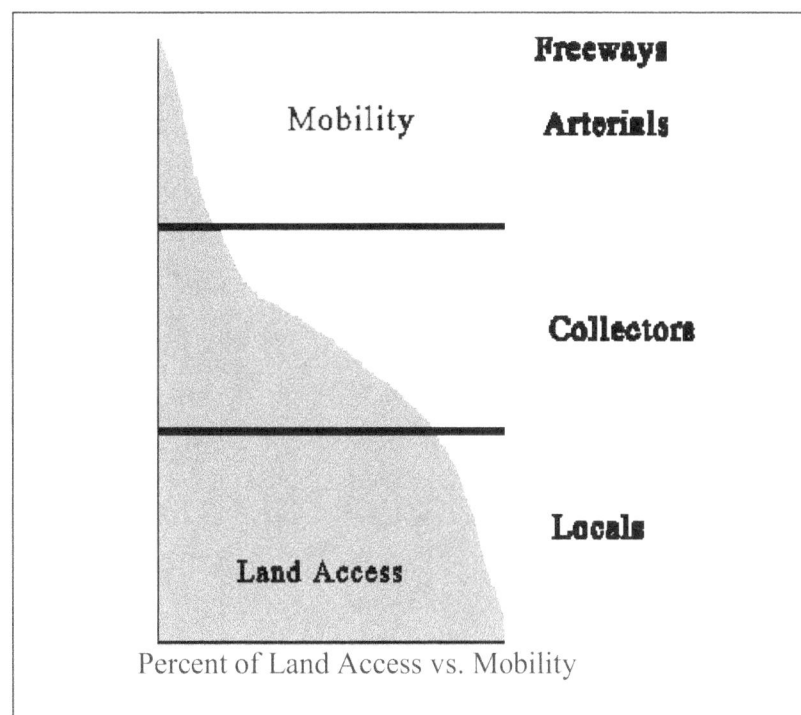

Source: FHWA guide on Functional Classification

Volpe National Transportation Center

Mobility and access determine functional class.

1. Demographic characteristics of area (urban vs., rural)

    2. Land uses adjacent to a route
    3. Route continuity
    4. Route ability to serve population centers and generators
    5. Trip length characteristics of a route
    6. Spacing of other routes with regard to function of the route

The functional classification of the portion of Route 2A from Route I-95 in Lexington to Route 2 at the Lincoln/Concord town line is a Minor Arterial. This classification is due to the connectivity to the highway system, the surrounding land uses, and a value judgment by the Massachusetts Highway Department in determining what type of trips the roadway serves. Changing the functional classification as a prelude to changing the design of Rte 2A can indirectly change the classification of neighboring roadways as well. Table 1 summarizes the roadway classes in the functional classification system.

Table 1: Functional Classification of Criterion

Class	Functional System	General Characteristics
Urban	Principal Arterial	1. Serve Major centers of activity, highest traffic volume corridors and longest trip desires. 2. Carry high proportion of total urban travel on minimum of mileage. 3. Integrated internally between major rural connections. 4. Carry major portion of trips entering and leaving urban area as well as through movements. 5. Serve significant intra-area travel. 6. Provide continuity of rural arterials that intercept urban boundary. 7. Service to adjoining land is subordinate to major travel movements.
	Urban Interstate	1. Principal arterial already designated as part of the interstate system.
	Urban Other Freeways	1. Non-Interstate, controlled access facilities, expressways
	Other Principal Arterials	1. Principal arterials without controlled access
	Minor Arterial	1. Interconnect with and augment principal arterial system
	Collector	1. Provides for land access and traffic circulation within residential neighborhoods, commercial or industrial areas 2. May penetrate residential neighborhoods. 3. Distribute trips from arterials to locals and collect trips from locals and channel them to arterials
	Local	1. Provide direct access to adjoining land. 2. Provide access to higher systems 3. Lowest level of mobility: discourages through traffic movement
Rural	Principal Arterials	1. Serves corridor having trip lengths and travel densities indicative of substantial statewide or interstate travel 2. Serve virtually all urban areas of 50,000 and over and large majority of those with population over 25,000. General, rural arterials penetrate the urban boundary 3. Provide an integrated network without stub connections
	Rural Interstate	1. Principal arterial already designated as part of the interstate system
	Other Principal Arterial	1. All non-interstate principal arterials
	Minor Arterial	1. Link cities and larger towns (or major resorts) and form an integrated network providing interstate and inter county service 2. Spaced at intervals so that all developed areas are within a reasonable distance of an arterial 3. Provide service to corridors with trip lengths and travel density greater than those served by rural collector or local systems. They should have high overall travel speed with minimum interference to through movements.
	Collector	1. Primarily serves intra-country travel. 2. Provide service to any county seat, larger towns, and other traffic generators not served by higher system. Example generators include consolidated schools, county parks, mining, agricultural areas, etc.
	Local	1. Serve primarily access to adjacent land. 2. Provide service over relatively short distances.

Volpe National Transportation Center

4.2.2 Functional Speed

The proposed design speed will affect the roadways dimensions more than anything else. The design speed is the highest speed at which a motorist can drive safely under ideal conditions, regardless of the posted speed limit. The design speed is more of a policy decision about how fast motorist should be encouraged to drive. AASHTO provides guidelines on speed, but in most cases the state will set the speed limit but may be altered by communities with the states approval. A standard handbook, the Manual on Uniform Traffic Control Devices recommends that to determine the speed that should be posted, a traffic engineering study should be done that considers the following factors:

- The nature and condition of the roadway surface and shoulders, the roads steepness, curves, and stopping sight distances.
- The 85th percentile speed, which is the speed that 85% of the traffic is traveling, regardless of the current speed limit.
- Land uses and activities that border the road and their compatibility with different speeds.
- Use of the road by pedestrians, bicyclists, and for parking.
- The types and number of accidents reported during a 12 month-period.
- The speed considered safe for the all other factors.

Rte 2A functional speed varies by time of day and location with respect to intersections, During AM and PM peak hours of congestion the speed varies between 20 and 30 mph. During the off-peak hours, it can average 40mph or more.

4.2.3 Traffic Volumes

The traffic volumes on the roadway are an important part of this analysis, when they occur, their magnitude, and what is the road's ability to handle that volume using it, are just a few of things considered. It is standard practice in transportation planning to look at the traffic volumes on the roadway now and what it might handle 10 to 20 years into the future based on some assumed growth rate that that area will have.

Special attractors like the park, the airport, and possible future development at Hanscom Field and the Air Force Base complicate traffic growth projections for Rte 2A ten to 20 years into the future. This variability is due to either development plans that are very sketchy or uses that can be seasonal or special event related can cause large variations in traffic volumes and patterns. All of the special attractors mentioned above contribute to this difficulty in determining how much traffic growth Rte 2A might see and how the roadway would handle it. Another issue is determining when it would occur, whether it be spread out during the whole day or spike in the in the AM and PM peak hours as it does now.

4.2.4 Design

Presently Route 2A is a two lane unlimited access roadway. The posted speed is 40 mph. There are no sidewalks or paved shoulders. Traffic flow controls are limited to a centerline and edge of road pavement markings. The road has both "Passing" and "No Passing" zones.

Land use along the north wayside is mostly the Minute Man National Historical Park. Land use along the south wayside and sections of the north wayside is sparse residential. Over time the National Park Service has been continually taking control of the residential land along the road in fee, by grants and other agreements.

Any alteration to 2A must meet the performance standards described below in order to allow Route 2A to function as a minor arterial. If alterations to Route 2A decrease its ability to function as a minor arterial, then alterations to regional roads may need to be made and mitigate any changes in traffic flows and patterns on neighboring roads.

The steps to select the design speed and geometric specifications for proposed alterations to Route 2A are as follows:
- Determine the proposed functional classification of the road.
- Determine the anticipated service volume and type of traffic.
- Select the level of service required to satisfy that function at the anticipated service volume.
- Select design speed based on the speed a driver is likely to expect and meet the LOS required for the selected functional classification. Traffic volumes may also impact the selection of design speed. With all other factors being equal, a higher volume highway may justify a higher design speed. However the designer should consider that at low volumes, drivers are likely to drive at high speed.

The American Association of State Highway Transportation Officials (AASHTO) provides guidance on what methods are used to analyze traffic problems and sets standards to help communities set goals for improving their transportation situation.

The three terms used to define traffic flow on roadway are the number of vehicles per hour, average traveling speed in miles per hour and traffic density in vehicles per mile. When reducing running speed by interference or other means, vehicles travel close together. This increases the traffic density. "When the interference becomes so great that despite the closer spacing and increase density the average speed drops below that necessary to maintain stable flow, there becomes a rapid decrease in speed and traffic flow, and severe congestion sets in." (A Policy on Geometric Design of Highways and Streets, 1984, AASHTO) Increases in density can exacerbate air and noise pollution, create unsafe conditions and negatively impact the visitor experience while driving to the park.

Alterations to Route 2A which lower design speed standards would increase traffic density at current volumes. Any alteration should ensure that the roadway operates at a Level of Service C, which is the minimal (worst) LOS recommended by the AASHTO for arterials in an urban or suburban setting.

Volpe National Transportation Center

4.3 Rte 2A Traffic Volumes: Historical, Current, and Projected

Route 2A has seen tremendous growth in the last 40 years. The growth can be examined by looking at how the volume rise and fall along it due to its interaction with adjoining roadways. A key consideration is to understand how the traffic fluctuates by time of day, by day of the week, and by month. This analysis culminates in observing how an average daily volume changes year by year. Please refer to Figure 4 for an understanding of the magnitude of how these increases have occurred over time on Route 2A.

There are four methods used to describe the volume. The first 2 methods add both directions of traffic while the last 2 methods identify each lane. The first is Average Daily Traffic (ADT). This represents an average day during an average month in a given year. The second method is an examination of the average weekday traffic volume (AWDT). AWDT is generally used to examine the impact of commuter and work trips along the roadway. It is not useful for examining weekend trips or trips that focus on recreational uses that might occur on the weekend. The third method focuses on examining a peak 3-hour time period, usually either AM or PM during the weekday. For Route 2A the AM peak period ranges from 7:00 AM to 10:00 AM and the PM period ranges from 4:00 PM to 7:00 PM. The fourth method looks at the peak hour on weekdays. The AM peak hour on Route 2A is between 8:00 AM and 9:00 AM while the PM peak hour is between.

The traffic volumes presented were derived from several sources, at various times, using different methodologies and equipment. The data presented represents an approximation of these data sources and have been adjusted slightly to compensate for problems with these variations. The sources include traffic reports from consultants hired by the towns of Lincoln, Lexington, and Concord. Massport and the Air Force Base performed several studies focusing on the Airport and Route 2A as well. State highway officials, regional planning agencies, and local governments were consulted as well. ADT volumes were collected in approximately 5-year increments from 1960 to 2000.

There are 6 locations that have been historically considered good places to conduct counts based on their proximity to major intersections with adjoining roads. The locations are Route 2 and Route 2A in Concord, located west of Crosby's corner. The second location is east of Crosby's corner on the Route 2A cut-off. The third location is East of Lexington Road. The fourth location is on Route 2A East of Hanscom Drive. The fifth location is West of Interstate 95 but East of Massachusetts Avenue. The last location is East of Interstate 95

4.3.1 Historical

The oldest traffic volumes that could be obtained date back to 1960. They show that
The traffic along Route 2A between Crosby's Corner and East of Route 128 was fairly constant, averaging between 5 and 7 thousand vehicles daily. This stretch of roadway shows very consistent growth of about 500 cars a year until 1970. Route 2A West of Interstate 95 then starts growing much faster than Route 2A, East of Interstate 95. Between 1970 and 1980 the roadway West of Interstate 95 sees a growth of about 1000 cars a year as opposed to points East of Interstate 95 due to either new employment or increase use of the Airport. This shows that Interstate 95 starts becoming a feeder to access locations in and around the Park that continue to the present.

Volpe National Transportation Center

Figure 4: Historical ADT on Rte 2A

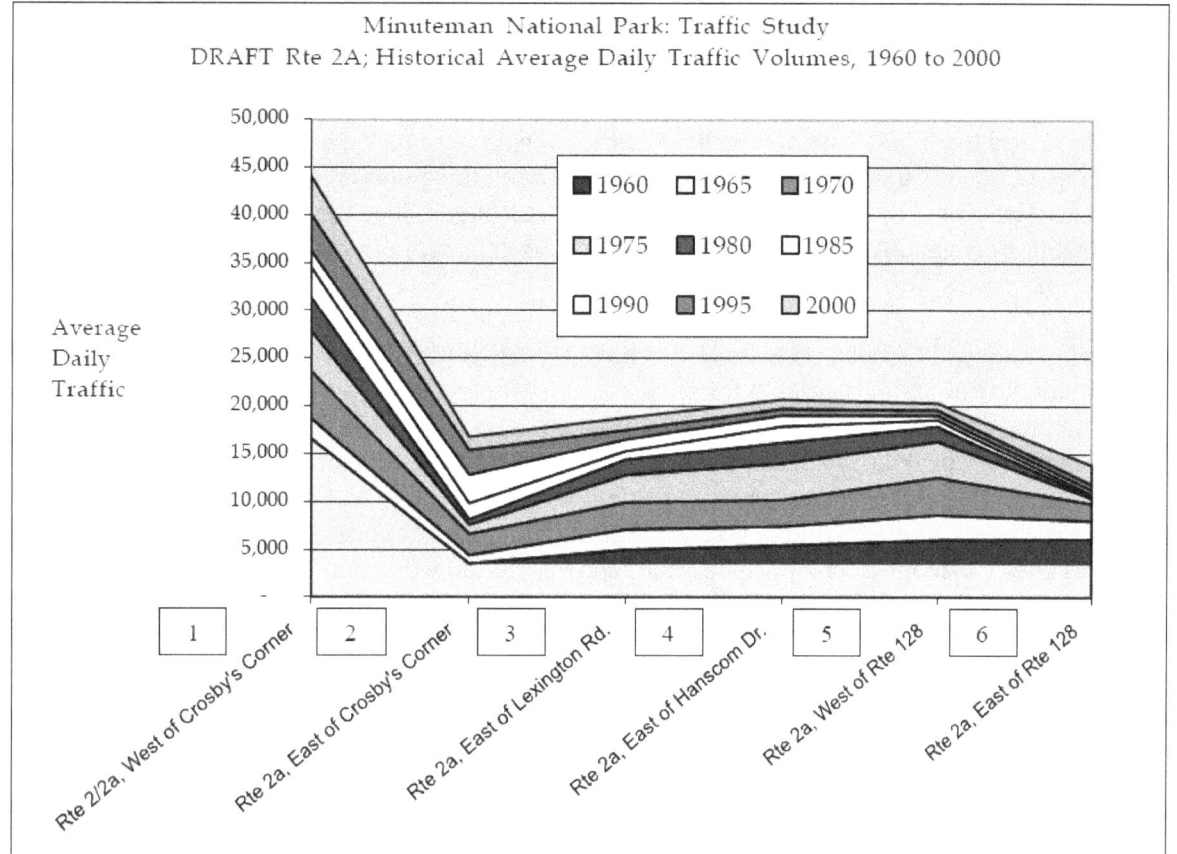

Source: Volpe National Transportation Systems Center. 2001

It is important to observe that the rate that the traffic along Route 2A grows starts decreasing after 1980, showing that either a saturation point is starting to be reached in the amount of traffic this roadway can handle is being reached or the population and employment in the area is declining. The demographic data doesn't show a decline so this possibility is doubtful but the idea of congestion on the Route 2A inducing people to seek alternate routes is a possibility that will be explored further in this report. Route 2A, West of Crosby's Corner merges with Route 2 at this point and it is evident from Figure 4 that the volumes at this point are significantly higher and not representative of the volumes around the Park.

The entrance to the Visitor Center is near Hanscom Drive and the volumes near this location have been historically the highest within the Park boundaries. This makes sense when one considers where the traffic is coming from and where it will go. The location of the Airport and employment sites in this area attracts a substantial number of trips daily. The aspect of trip patterns is vital in understanding why the traffic is highest at this point will be discussed in more detail later in the report.

Volpe National Transportation Center

4.3.2 Current

The current conditions show little growth on a yearly basis when compared to the historical trends of this area. Route 2A experiences about 20,000 vehicles daily. With some peaking of trips around Hanscom Drive but not as significantly as in previous years. The traffic drops of to 15,000 vehicles East of Interstate 95 and to 18,000 East of Crosby's Corner. Again we can see a substantial spike when it merges with Route 2, West of Crosby's Corner. The traffic at this point is more than twice the traffic on Route 2A near Hanscom Drive. It is important to note that Route 2 does have 1 more lane in each direction then does Route 2A, so this means that Route 2A carries about the same level of traffic per lane as Route 2.

There have been several recent studies done that have looked at traffic levels along Route 2A by time of day and by month that wasn't available for the historical analysis in the previous section. In Figure 5 we can observe how traffic varies by time of day for a location on Route 2A, East of Hanscom Drive. This shows that in the AM peak hour approximately 1,150 vehicles head eastbound while only 750 travel westbound. The midday levels drop quickly to about halve of the AM peak hour volumes, at about 500 to 600 cars per hour. It is interesting to note that the midday sees about the same amount of traffic eastbound as goes westbound. The PM peak hour shows that the westbound volumes increase to about 1,150, the exact mirror image of the AM levels with the eastbound levels being a little more than halve of the westbound, at about 750 vehicles per hour. The PM peak period seems to last longer than the AM peak period and tapers off to just a couple hundred vehicle per hour after 10:00 PM. Figure 6 and 7 show how the peak hour volume changes by the peak direction of flow for the six locations that were described in the previous section.

Volpe National Transportation Center

Figure 5: Daily Variation of Volume on Rte 2A

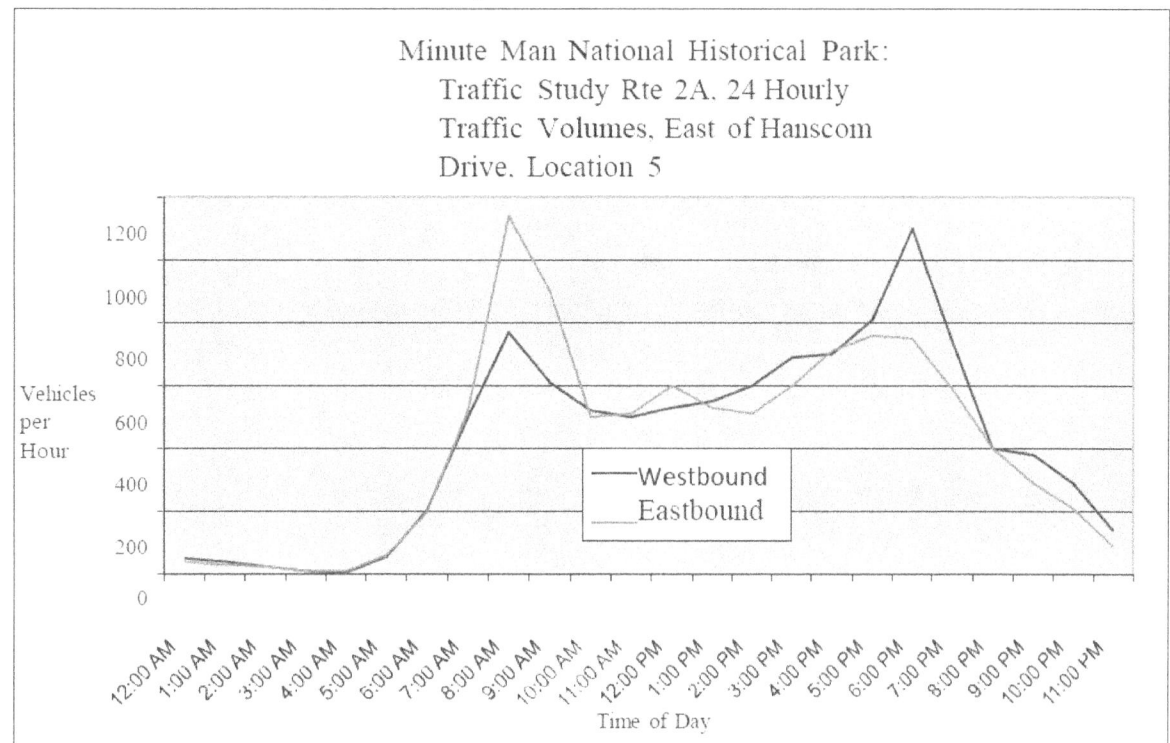

Source: Volpe National Transportation Systems Center, 2001

Figure 6: AM Peak Hour Volumes on Rte 2A

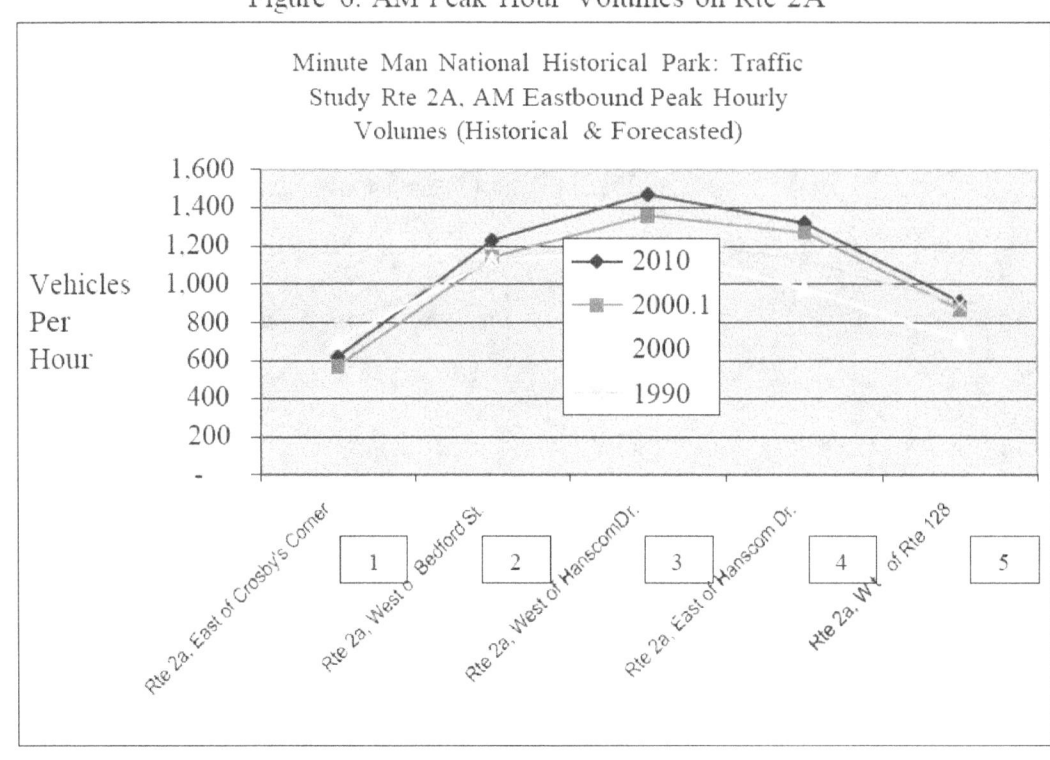

Source: Volpe National Transportation Systems Center. 2001

Figure 7: PM Peak Hour Volumes on Rte 2A

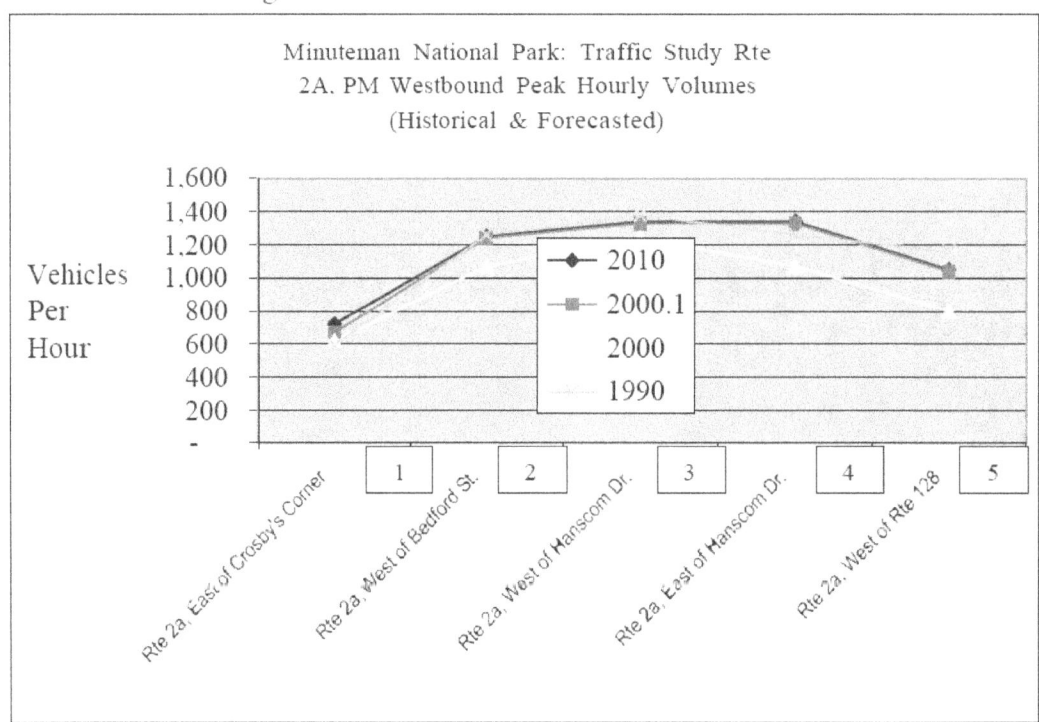

Volpe National Transportation Center

The monthly variation tells another story about how traffic varies by season. The traffic levels peak in the summer months, namely August and in October as well. The increase in October corresponds to the foliage season. The traffic on Rte 2 bottoms out during the winter months, namely December, and January. Spring months are generally about 5% less than summer months.

4.3.3 Projections

Projecting traffic growth is very difficult to do and many assumption about land use and demographic patterns in the area need to be made. The research on projections is derived from two sources, Massport and the Boston Metropolitan Planning Organization. In both cases the demographic assumptions are similar and show about 10% growth in the area. The employment in and around the park should increase slightly but not pose a significant problem. The problem is determining how much activity and or development will occur at the Airport over the next 10 years. The Airport refers to its uses as a Regional Airport as well as an air base. Will the Air Force Base close? Will it be developed? Will Massport increase its use of the Regional Airport? Massport assumes limited growth in the commercial traffic and there is a strong lobbing effort underway to make sure the Air Force Base remains open. If these things occur the traffic growth will be less than 1% a year along Route 2A over the next ten years. If the base closes and the land is developed or Massport increases it operations above what was projected in the 1997 GEIR, substantial increases in traffic could occur

4.4 Trip Patterns: Origins, Destinations, & Flows

Trip patterns help us understand where vehicle come from, the origins, where they go to, the destination, and what routes the travel on to go between them, the flows. This is important because once we understand these patterns we can help promote changes in land use, demographics, and roadway configuration that can impact the roadway in question positively.

4.4.1 Origins

The origins are mainly residences or sites that might be an intermediate point for future travels, like the Park's visitor center. Six towns were considered in the analysis; Acton, Bedford, Burlington, Concord, Lexington, and Lincoln. The unit of measurement was households that were derived from Census information and the Metropolitan Area Planning Council (MAPC) forecasts. Please refer to Table 2 for a comparison of growth by town and by year. Lexington contains the largest number of households and is expected to see minimal growth. The other communities on average are expected to see between 12% and 47% increase in the number of their households between 2000 and 2020. Lincoln, the town that the majority of Route 2A lies within in this study has the smallest amount of households, with only 2,930 in 2000 but experience 27% growth, representative of the study area. If each new household in Lexington alone generates 10 vehicle trips a day, that translates into over 30,000 vehicles traveling the roadways in and around Minute Man National Historical Park in 2020.

The Park acts like an origin and a destination in this analysis. Based on the National Park Service Visitation Data for 2001, the Park saw over 1 million visitors to the Minuteman Nation Park. The problem with this information is determining what fraction of these actually visited the Visitor Center and the parking lots along Battle Road as opposed to touring the sites in Concord. This number is only expected to increase with time. Even at a modest 1% growth rate a year, that is over 220,000 more visits by 2020 using Rte 2A and Lexington Road as the primary means of access to the park.

4.4.2 Destination

Destinations can include employment sites, retail establishments, schools, and landmarks. Employment is probably makes up the largest proportion of destination in this study. As we can see from Table 3 Bedford has the greatest number of people employed within it with 24,600. Lexington is second 19,400 respectively. The employment growth is less than the households, averaging between 2% and 37% between 1990 and 2020. The greatest growth occurs in Concord with the 37%.

Table 2

Household Growth by Community

	Households			Growth	
Town	1990	2000	2020	90 to 20	90 to 20
Acton	6,600	7,700	9,700	3,100	47%
Bedford	4,500	4,400	5,000	500	12%
Concord	5,700	5,800	6,600	900	31%
Lexington	10,500	12,000	13,600	3,100	29%
Lincoln	2,600	3,000	3,700	1,100	39%

Source MAGIC Phase1 Report CTPS 2000

Employment Growth by Community

	Households			Growth	
Town	1990	2010	2020	90 to 20	90 to 20
Acton	9,400	11,300	11,900	2,500	27%
Bedford	24,600	23,500	25,200	500	2%
Concord	11,600	14,500	16,000	4,400	37%
Lexington	19,400	21,600	22,700	3,300	17%
Lincoln	1,700	1,700	1,900	200	10%

Source MAGIC Phase1 Report CTPS 2000

Retail has minimal impact on the area. The Minuteman Vocational School is located on Route 2A between Airport Road and Interstate 95. The trips that this school attracts in relation to the total number of trips on Route 2A are minimal as well. The Visitor Center attracts on average 60 to 80 cars a day with the satellite parking locations attracting between 40 to 60 vehicles combined on a peak daily average.

4.4.3 Flows

Analyzing turning movements from several traffic studies and comparing them with the origin and destination data were used to help determine the flows and travel patterns in the area.
The analysis will look at each travel lane by peak hour at three different locations.

Volpe National Transportation Center

The first location is on Route 2A, East of Lexington Road. Please refer to Figure 8 for a graphical display of where the flows come from and where they go for the AM. This shows that 1,130 vehicles travel eastbound in the peak AM hour. Of these 42% come from Lexington Road, and 54% come from Route 2, West of Crosby's Corner. Only 4% use the Concord Turnpike. In the westbound direction there are only 480 vehicles in the peak AM hour with the 54% heading for Route 2 and 48% heading for Concord using Lexington Road. Please refer to Figure 9 for a graphical display of where the flows come from and where they go in the PM. The westbound direction has 1,370 vehicles with 60% heading towards Concord via Lexington Road and 39% heading west on Route 2. In the eastbound direct there are 590 vehicles with 58% coming from Route 2 and 41% coming from Lexington Road.

The second location is on Route 2A, East of Hanscom Drive. Please refer to Figure 10 for a graphical display of where the flows come from and where they go for the AM. This shows that 1,120 vehicles travel eastbound in the peak AM hour. Of these 94% come from points Route 2A west of Hanscom Drive. 6% head south from Hanscom Drive and make a left hand turn. 99% of these trips continue down Route 2A to Interstate 95. In the westbound direction there are 720 vehicles in the peak AM hour with the 66% heading to points along Route 2A west of Hanscom Drive. 34% of the traffic turns left onto Hanscom Drive and heads north. 99% of these trips come from Route 2A and Interstate 95. Please refer to Figure 11 for a graphical display of where the flows come from and where they go in the PM. The westbound direction has 1,280 vehicles with 88% continue past Hanscom Drive on Route 2A. 12% of the vehicles heading north on Hanscom Drive come from Route 2A. In the eastbound direction there are 750 vehicles with 70% coming from Route 2A, west of Hanscom Drive and 30% coming from Hanscom Drive.

The third location was Hanscom Drive, just north of Rte 2A. In the AM peak hour, 60% of all of the trips to Hanscom Airport and the surrounding uses arrived via Rte 2A from the east. The remaining 40% came from the west on Rte 2A. The majority of trips were inbound during this time, 700 compared to only 190 vehicles heading south on Hanscom Drive in the morning. The destinations were equally divided between eastbound and westbound locations. In the PM peak hour for this location the majority of the vehicles were leaving, 670 as opposed to entering, 290. The majority of the vehicles, 68%, that were heading south down Hanscom Drive turned left, heading eastbound towards Rte 128. The trips entering were predominantly from this direction as well.

4.5 Analysis of Rte 2A Traffic

The analysis will focus on five criteria. They are volume to capacity ratios as a means to compare how their design compares with their function. The intersection level of service will rank intersection based on their efficiency to handle turning movements. Research on pedestrian delay in crossing the street as a function of traffic volumes will be used to determine how the traffic impacts pedestrians. Excess speeds or abnormally slow speeds can disrupt visitors to the park, help create safety hazards, and increase noise levels so an examination of them will be undertaken as well. Accident data in unison with speed information and the level of service can help identify potential safety problems along the roadway and its impact on the visitor to the park. The last item to consider is to examine what are the levels of noise in and around the park and to what extent they impact the visitor's experience.

Figure 8: Rte 2A Traffic Flows, AM Peak Hour, West of Hanscom Drive

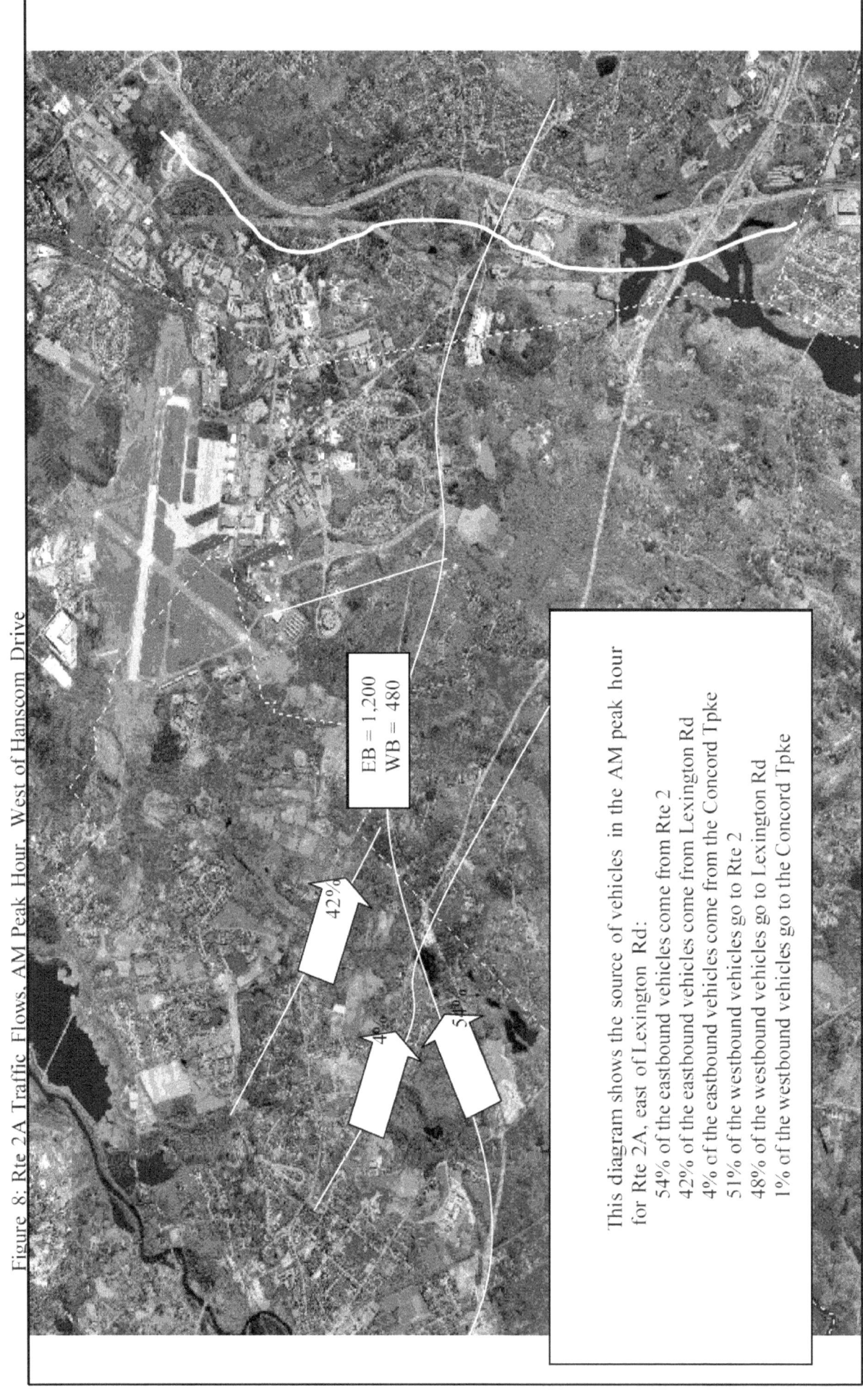

This diagram shows the source of vehicles in the AM peak hour for Rte 2A, east of Lexington Rd:
54% of the eastbound vehicles come from Rte 2
42% of the eastbound vehicles come from Lexington Rd
4% of the eastbound vehicles come from the Concord Tpke
51% of the westbound vehicles go to Rte 2
48% of the westbound vehicles go to Lexington Rd
1% of the westbound vehicles go to the Concord Tpke

Volpe National Transportation Center

Figure 9: Rte 2A Traffic Flows, PM Peak Hour, West of Hanscom Drive

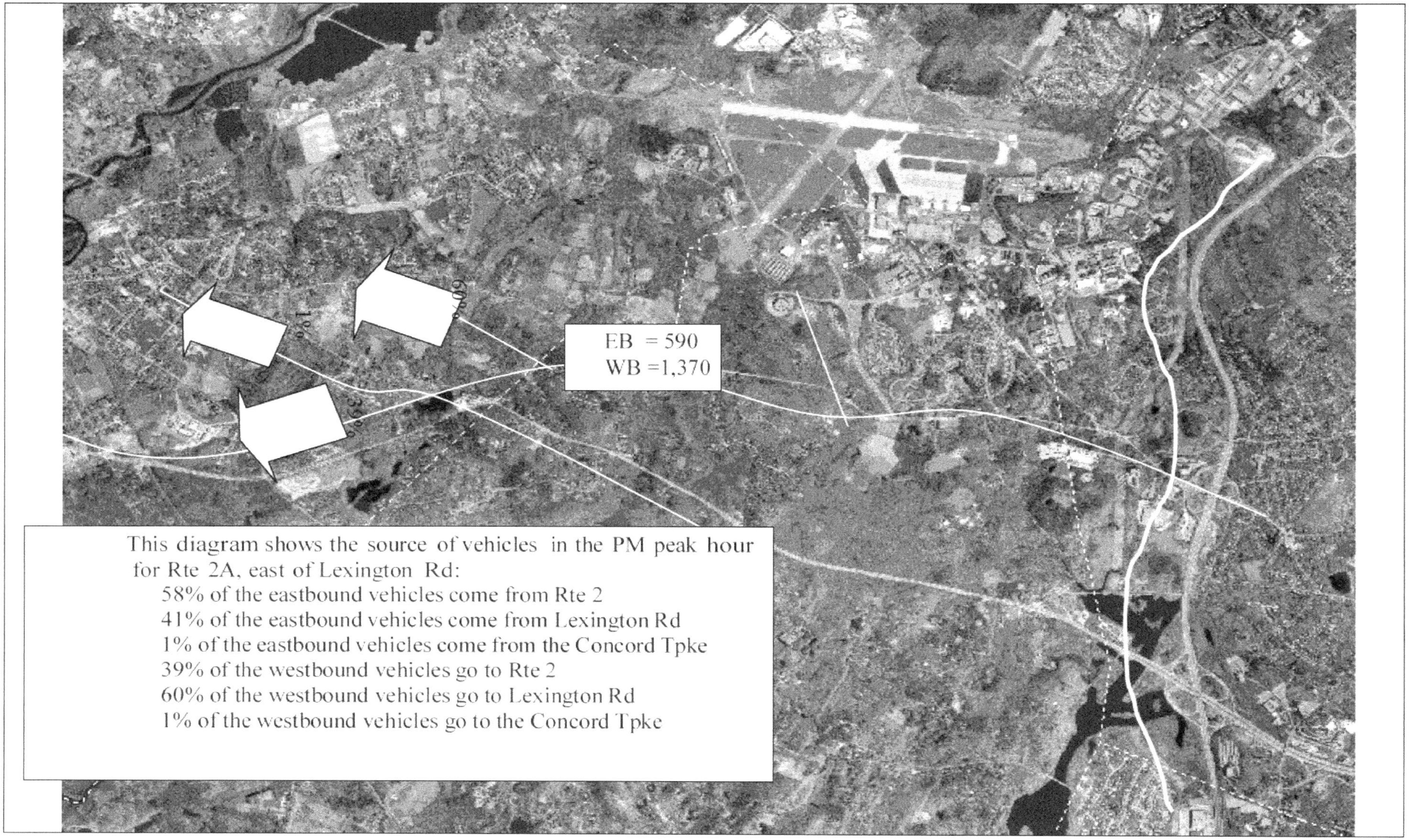

EB = 590
WB = 1,370

60%
39%
1%

This diagram shows the source of vehicles in the PM peak hour for Rte 2A, east of Lexington Rd:
58% of the eastbound vehicles come from Rte 2
41% of the eastbound vehicles come from Lexington Rd
1% of the eastbound vehicles come from the Concord Tpke
39% of the westbound vehicles go to Rte 2
60% of the westbound vehicles go to Lexington Rd
1% of the westbound vehicles go to the Concord Tpke

Volpe National Transportation Center

Figure 10: Rte 2A Traffic Flows, AM Peak Hour, East of Hanscom Drive

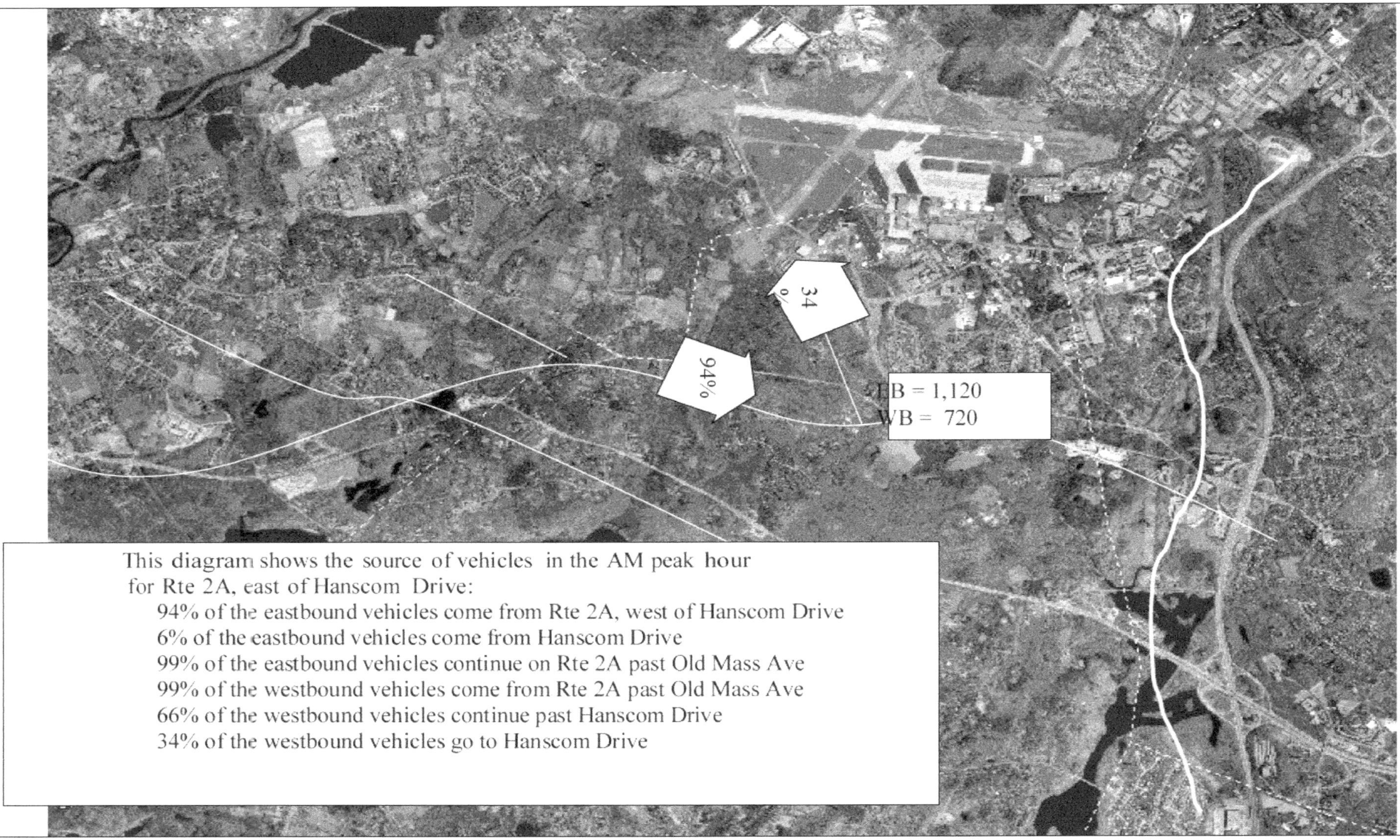

EB = 1,120
WB = 720

34%

94%

This diagram shows the source of vehicles in the AM peak hour for Rte 2A, east of Hanscom Drive:

94% of the eastbound vehicles come from Rte 2A, west of Hanscom Drive
6% of the eastbound vehicles come from Hanscom Drive
99% of the eastbound vehicles continue on Rte 2A past Old Mass Ave
99% of the westbound vehicles come from Rte 2A past Old Mass Ave
66% of the westbound vehicles continue past Hanscom Drive
34% of the westbound vehicles go to Hanscom Drive

Volpe National Transportation Center

Figure 11: Rte 2A Traffic Flows, PM Peak Hour, East of Hanscom Drive

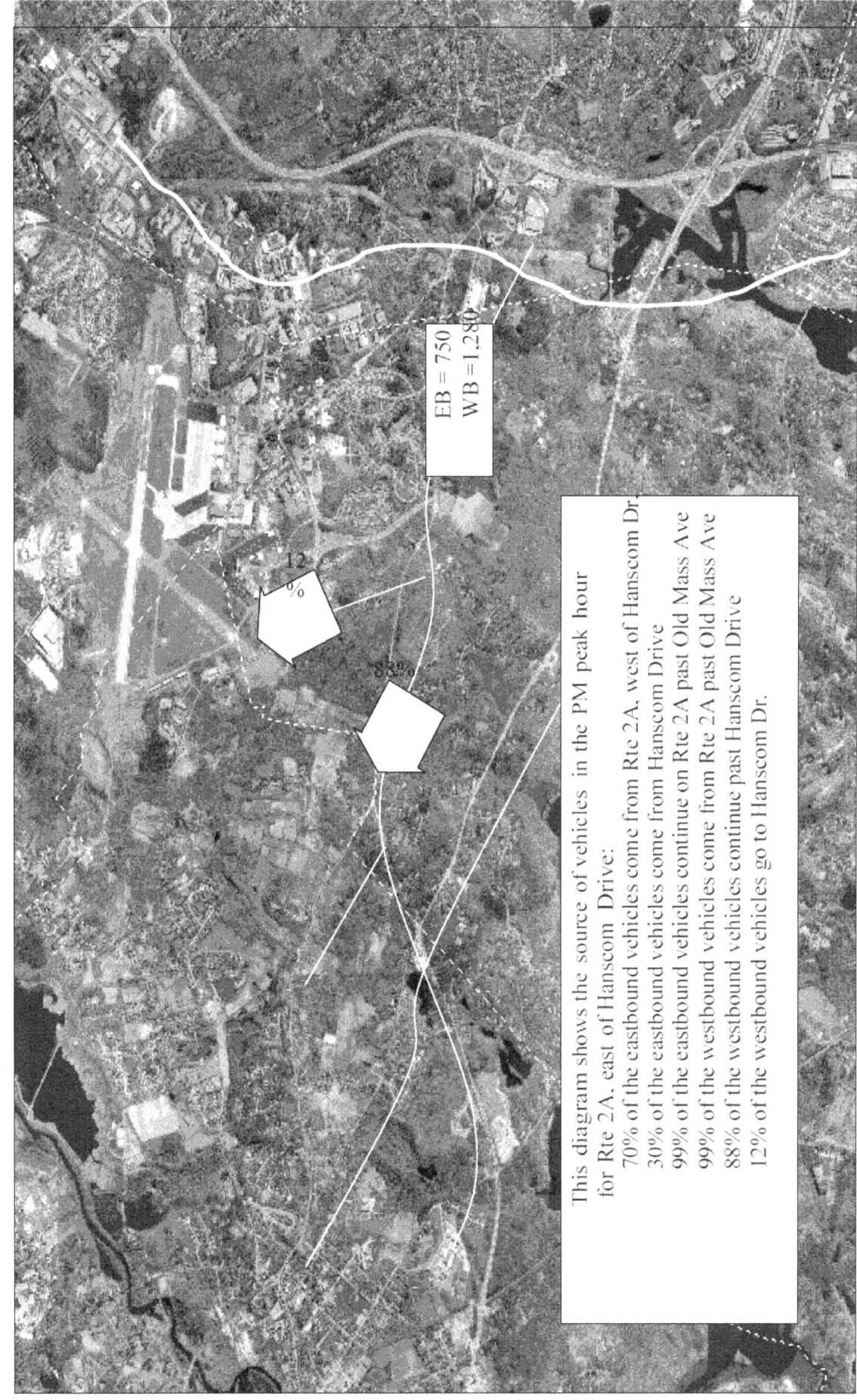

Volpe National Transportation Center

Figure 12: Hanscom Drive Traffic Flows, AM Peak Hour, North of Rte 2A

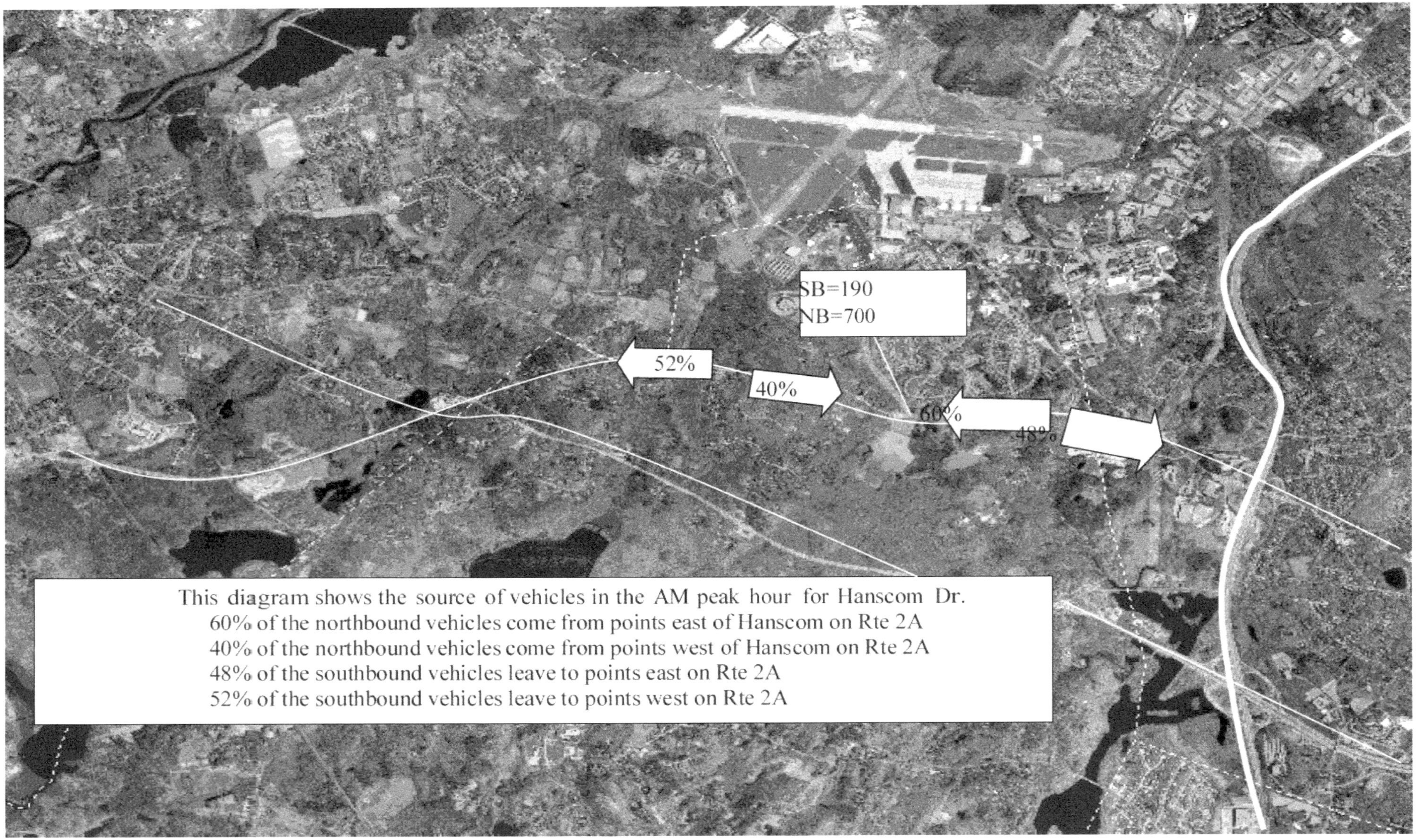

SB=190
NB=700

60%
40%
52%

This diagram shows the source of vehicles in the AM peak hour for Hanscom Dr.
60% of the northbound vehicles come from points east of Hanscom on Rte 2A
40% of the northbound vehicles come from points west of Hanscom on Rte 2A
48% of the southbound vehicles leave to points east on Rte 2A
52% of the southbound vehicles leave to points west on Rte 2A

Volpe National Transportation Center

Figure 13: Hanscom Drive Traffic Flows, PM Peak Hour, North of Rte 2A

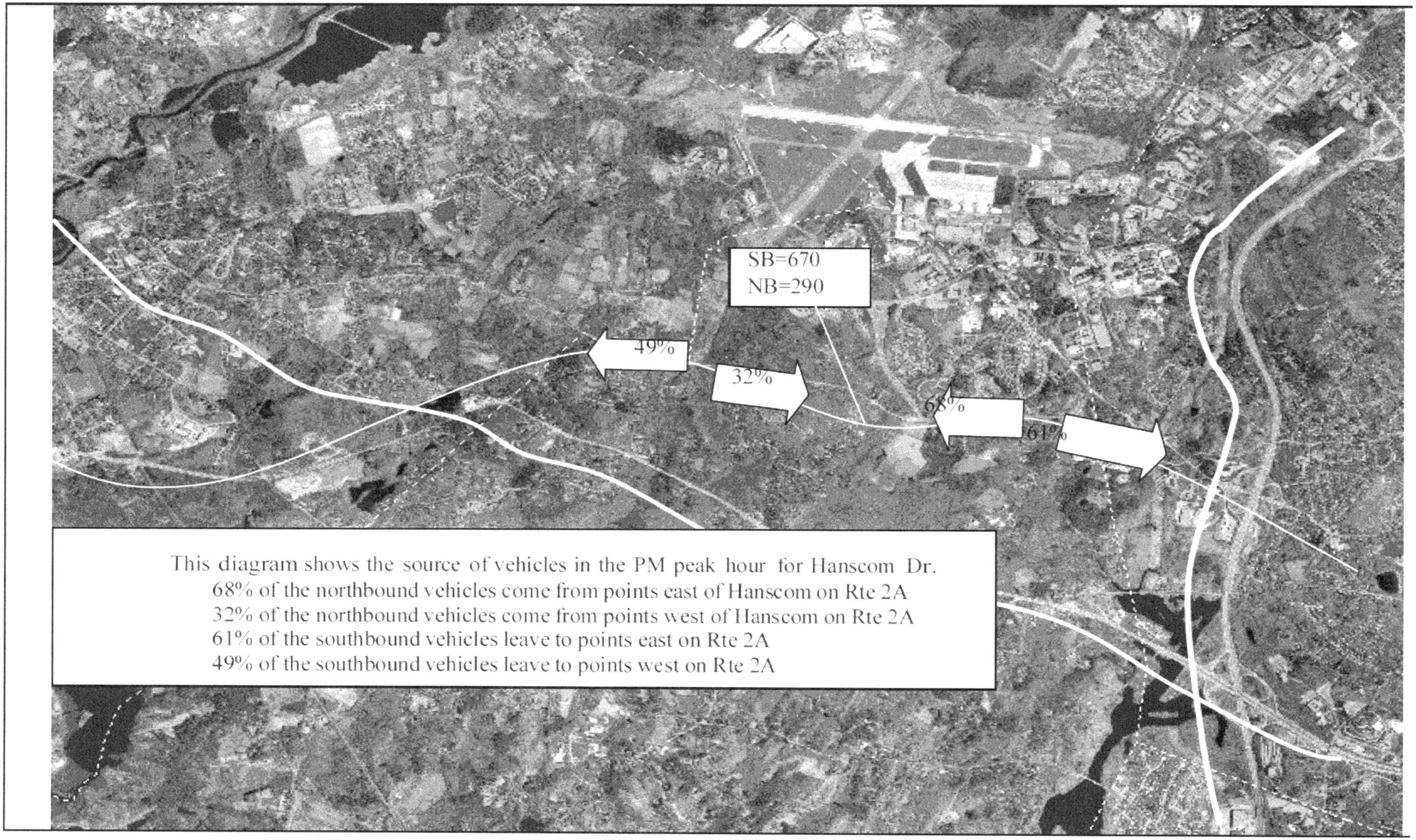

SB=670
NB=290

32%
49%
68%
61%

This diagram shows the source of vehicles in the PM peak hour for Hanscom Dr.
68% of the northbound vehicles come from points east of Hanscom on Rte 2A
32% of the northbound vehicles come from points west of Hanscom on Rte 2A
61% of the southbound vehicles leave to points east on Rte 2A
49% of the southbound vehicles leave to points west on Rte 2A

Volpe National Transportation Center

4.5.1 Volume to Capacity

Volume to capacity ratios (V/C) are a measure to quantify how well the road can handle the traffic on it versus what it was meant to do. This ratio can use daily or peak hour volumes but in this report we will look at peak hour volumes as more representative of the worst case conditions experienced in the AM and PM peak hours. The closer to one the value is, the greater the congestion levels. The design capacity for Route 2A on average is about 1,500 vehicles per hour per lane. The maximum volume on Route 2A being experienced in the AM peak hour is 1,130.

Location: Route 2A, west of Hanscom Drive: eastbound direction, AM peak hour
Volume / capacity = v/c   1,130/1,500 = 0.753

Location: Route 2A, west of Hanscom Drive: Westbound direction, PM peak hour
Volume / capacity = v/c   1,370/1,500 = 0.913

Based on these two locations, it appears that the level of congestion is reaching a point where the roadway can't effectively handle any more traffic, especially in the PM peak hour westbound

4.5.2 Intersection (Level of Service)

Intersection Level of Service (LOS) is used to determine how efficiently an intersection is operating by examining factors such as speed travel time, interruption, freedom to maneuver, driver comfort, convenience, safety, and operating costs. The ranking is based on a letter grade to identify the degree to which these factors have been satisfied. The LOS scale goes from A, the best service to F, the worst. Please refer to Table 5 for a description of each grade. These rankings help us determine what intersection need improvement and those that don't based on the minimal LOS that should occur at any given intersection. Please refer to Table 6 for an assessment of what is the minimal LOS that is considered acceptable.

Table 6 presents the relationship between highway type and location and the level of service appropriate for design, suggested by the AASHTO Green Book. Taking into consideration specific traffic and environmental conditions, the responsible highway agency should attempt to provide a reasonable and cost effective level of service. While the Highway Capacity Manual provides the analytical basis for design calculations and decisions, judgment must be used in the selection of the appropriate level of service for the facility under study. Once a level of service has been selected, all elements of the roadway should be designed consistently to that level. Rte 2A falls into the arterial roadway that is in an urban/suburban environment category and suggests a minimal LOS of C for all of its intersections.
Source: http://www.fhwa.dot.gov/environment/flex/ch04.htm

The United States Air Force Hanscom Base conducted a Traffic Impact Study in 2000 to determine how relocating 1000 base personnel to another site in Bedford would impact traffic. The results of this analysis showed several intersections with a LOS of C or worse in the base year, 2000. They identified problem locations with and without signalization. Several future alternatives were examined and due to the background growth in traffic and higher utilization of the surrounding land uses, these LOS only degraded with time unless some form of corrective action was taken. Table seven presents some sample locations for comparison.

Table 5: LOS Scale

Level of Service	Description
A	Free flow with low volumes and high speeds.
B	Reasonably free flow, but speeds beginning to be restricted by traffic conditions.
C	In stable flow zone, but most drivers are restricted in the freedom to select their own speeds.
D	Approaching unstable flow; drivers have little freedom to select their own speeds.
E	Unstable flow; may be short stoppages
F	Unacceptable congestion; stop-and-go; forced flow.

Source: Adapted from the AASHTO Green Book.[1] 1995 Highway Capacity Manual (Special Report 209), Transportation Research Board, Washington, DC, Third Edition, updated 1994

Table 6: Guidance on Minimal Acceptable LOS Levels

	Type of Area and Appropriate Level of Service			
Highway Type	Rural Level	Rural Rolling	Rural Mountainous	Urban and Suburban
Freeway	B	B	C	C
Arterial	B	B	C	C
Collector	C	C	D	D
Local	D	D	D	D

Source: Adapted from the AASHTO Green Book

Table 7: Sample LOS and Their Location

Location	Level of Service	
	AM	PM
Rte 2a at Hanscom Drive	F	F
Rte 2a at Bedford Road	D	F
Rte 2a at Lexington Road	F	B
Hanscom Drive & Old Bedford Road	A	A

Source: 2000 Hanscom Air Force Traffic Impact Report

4.5.4 Pedestrian Access

Another aspect of traffic volume is its impact on pedestrians' ability to cross the street in a timely manner. Research has produced quantifiable numbers on how long a person has to wait to cross the street given a certain volume of traffic on the roadway in a given time period. Rte 2a has averages about 1200 cars in the AM peak hour. This translates into a wait of several minutes based on Figure 14 graphical comparison of the functional relationship between volume and time to cross a street.

Figure 14: Impact of Traffic on Pedestrians

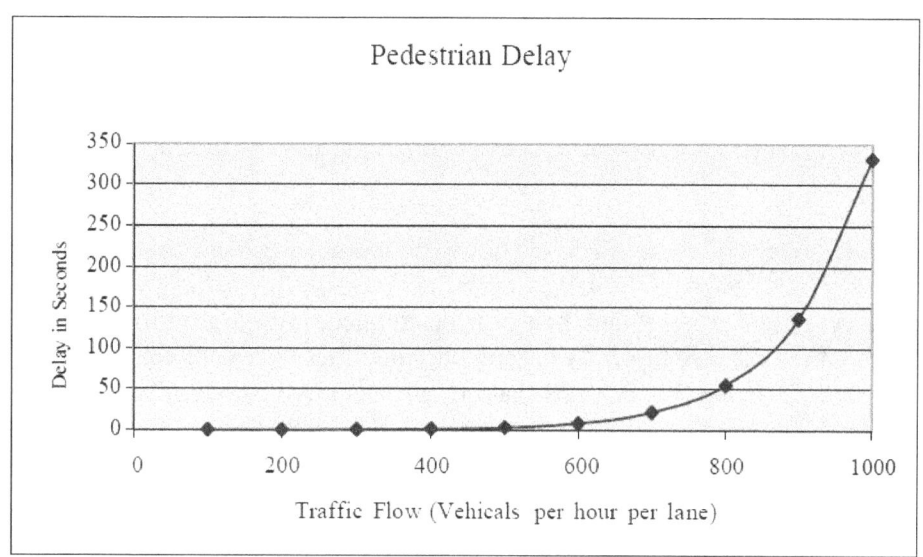

Source: David Spiller. Volpe Center 2001

4.5.4 Speed Issues

Another issue to consider is how does the speed limit relate to the roadway design, mobility issue, access to surrounding land uses, and pedestrian crossings. Based on these issues Rte 2A needs to re-access its speed limit of 40 mph. The 40 mph speed is appropriate for its functional designation and promotes mobility but doesn't consider safety issues relating to access to the surrounding land uses. Higher speeds reduce stopping distance that can impact other vehicles, pedestrians, and bicyclists alike. As Figure 15 shows that a speed of 37 mph requires about 250 feet for a small sedan to stop safely. If the speed is reduced to 19 mph, the stopping distance has reduced to 90 feet, a substantial improvement in safety for it and other modes as well.

Figure 15: Stopping Distances

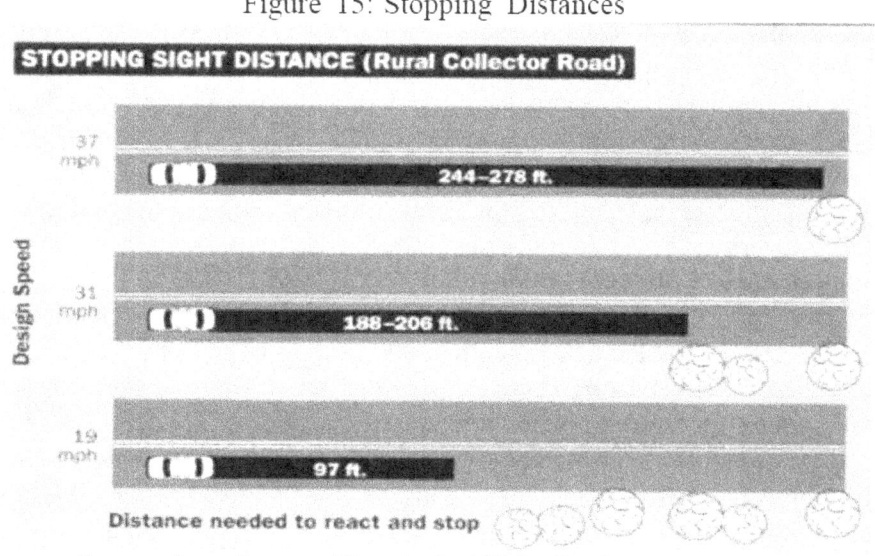

Source: http://www.clf.org/pubs/ (Take Back Your Streets)

In Massachusetts a municipality may set the speed limit for the roadway within its control but a "Traffic & Engineering Study" that has the approval from the Massachusetts Highway Department (MHD) and the Registry of Motor Vehicles must support it. While MHD may consider many factors when deciding wither the suggested speed limit is appropriate or not, they generally focus on the 85th percentile speed, which is the speed that 85% of the traffic is traveling, regardless of the current speed limit.

4.5.5 Accident & Safety

Highway engineers usually assert that their projects are necessary to improve "safety." While they are typically sincere, they are also typically relying on guidelines that were not developed with the goal of making roads safer for everyone. Projects that bring roads up to "modern" standards generally do not provide safe places for people to walk or bicycle-in part because they tend to increase traffic speeds-or they improve walking and bicycling facilities only as an afterthought. With better information, projects can and should be designed on the basis of a more comprehensive approach to safety.

Regardless of posted speed limits, motorists will drive faster when given the "safety cushion" of a wider road and greater sight distances. Higher design-speed roads have an insidious psychological effect on most motorists, prompting them to increase their speed unwittingly. When motorists drive faster, pedestrian accidents are both more likely and more serious.

The likelihood that a pedestrian will be hit increases at higher speeds because a motorist's ability to take in the surrounding environment is more limited. As we can see in Figure 16, a speed of 30 miles per hour, a motorist has a field of vision ("peripheral vision angle") spanning approximately 150 degrees, and will fix his or her vision about 1,000 feet ahead. At 60 miles per hour, the field of vision reduces to 50 degrees and the motorist will fix his vision at 2,000 feet. What this means in daily life is that a motorists driving at 25 m.p.h. or faster have difficulty perceiving that a pedestrian is ready to cross a street, deciding to slow down, and actually doing so. The normal driver usually decides to speed up, assuming that another car will stop. Source: http://www.clf.org/pubs/ (Take Back Your Streets).

Thus, from the point of view of pedestrian safety, widening a roadway is counterproductive. Figure 17 shows the probability of a pedestrian being killed is 3.5 percent more likely when a vehicle is traveling at 15 miles per hour. This increases more than tenfold to 37 percent at 31 miles per hour and increases to 83 percent at 44 miles per hour.[8] Pedestrian injuries also increase in severity with vehicle speed. As a 1994 treatise puts it, an injury's severity "depends primarily on the car's speed at impact with the pedestrian." The treatise ranks injuries on a scale of one (no injury) to six (fatality), and states that, in general, injury severity is 1.5 at 20 miles per hour, four at 30 miles per hour, and six at speeds greater than 35-40 miles per hour.

Accident data collected by the town of Lincoln in 2002 shows over 32 traffic accidents along Rte 2A. The majority of these, twenty occurred between Bedford Rd. & Hanscom Dr. These accidents generally fall into 2 basic groups, rear-end collisions and one person turning in front of someone else before they had a chance to stop. The cause for these varies between excessive speed, not paying attention, and miscommunication between drivers at intersections. As the traffic volumes, pedestrian, and bicyclist use along Rte 2A, Battle Road Trail, and Minute Man National Historical Park increase, the probability of accidents will increase as well in future years.

Figure 16: Ability to See Pedestrian Traffic

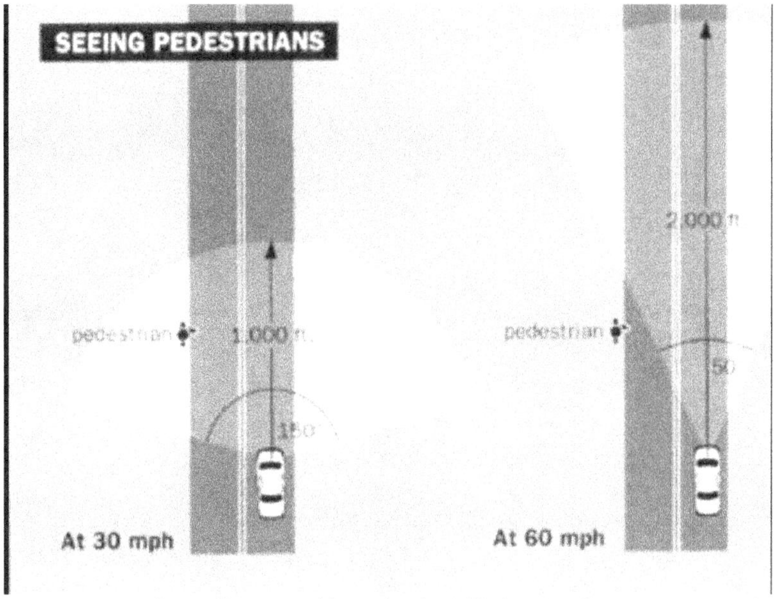

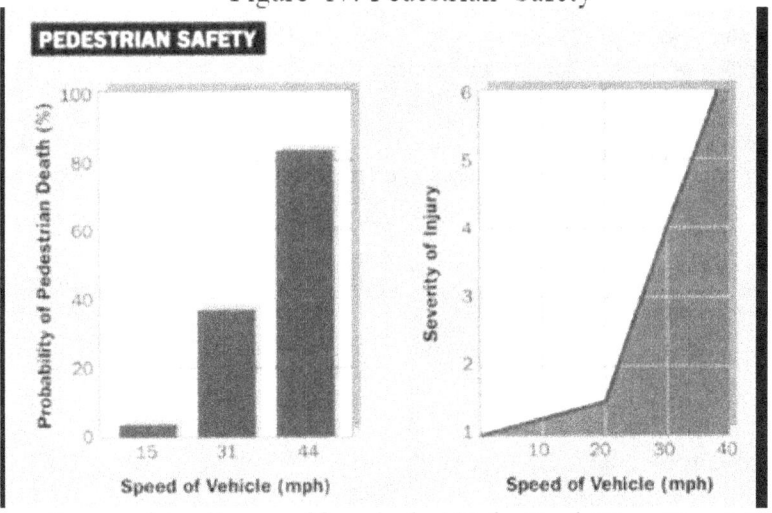

Source: http://www.clf.org/pubs/ (Take Back Your Streets)

Figure 17: Pedestrian Safety

Source: http://www.clf.org/pubs/ (Take Back Your Streets)

4.5.6 Noise Levels

Traffic noise can be annoying and harmful to those living, working, playing, or learning close to a roadway. Most studies have concluded that the threshold level of noise below, which no damage will occur, is 55 decibels (dBA). This level is lower than the noise levels deemed acceptable under federal regulations for places such as picnic areas, recreation areas, playgrounds, and parks, and for the exterior parts of residences, schools, and churches. The impact of any road project on noise levels should be carefully considered. Source: http://www.clf.org/pubs/ (Take Back Your Streets)

The effects of noise on people can take many forms, including the following.
- Hearing loss
- General health
- Classroom learning interference
- Sleep interference
- Communication interference
- Annoyance

Road projects can increase noise levels in a number of ways. As traffic speed increases, noise levels increase. From a distance of 16 yards, a car traveling at 31 miles per hour makes one-tenth as much noise as it would traveling at 56 miles per hour. Acceleration greatly increases noise levels at a particular instant, creating "noise peaks." Wider roads make acceleration by motorists more likely-for example, to return to higher travel speeds after stopping for a pedestrian-thereby increasing noise levels.

A government study of traffic calming in German neighborhoods has found that a reduction in average speed from 25 miles per hour to 12 miles per hour on residential streets was accompanied by a 14-decibel reduction in noise. This means that, on average, noise dropped to less than one-tenth its initial level. Other sources indicate that at somewhat higher speeds, an increase in traffic speed of 12-15 miles per hour corresponds to an increase in overall noise levels of 4-5 decibels.

Within Minute Man National Historical Park there are three main noise generators
- Aircraft – sporadic but loud, produce 100 db(a)
- Auto – peak hour congested volume, produce 70 db(a)
- Homes and businesses, steady around 30 db(a)

The majority of the Battle Road Trail is set back 50 feet or more from the road and is shielded from auto noise by tree growth during the summer. The noise shielding from trees does diminish in the fall and winter but also corresponds with diminished activity in the park.

There are several sections along the trail areas where traffic is more visible and audible, they are:
- Around Hanscom Dr.
- Near Airport Rd and Mass Avenue
- East of Lexington Road
- Around the parking lots
- Near the Visitor Center

An Environmental Impact Statement for a Landing Field in Juneau, Alaska conducted a study on 2001 on annoyance levels associated with airports on local neighborhoods. Its findings are shown in Figure 16. Other research by Lawrence Dallam in his Environmental Capacity of Neighborhood Streets equates traffic volume levels to specific noise threshold levels determined in the FHWA Traffic Noise Prediction Model. These results, as shown in Table 8, show that in a suburban setting such as Minute Man National Historical Park, an average daytime traffic volume of 750 vehicles per hour is considered acceptable.

Figure 18: Noise Annoyance Levels

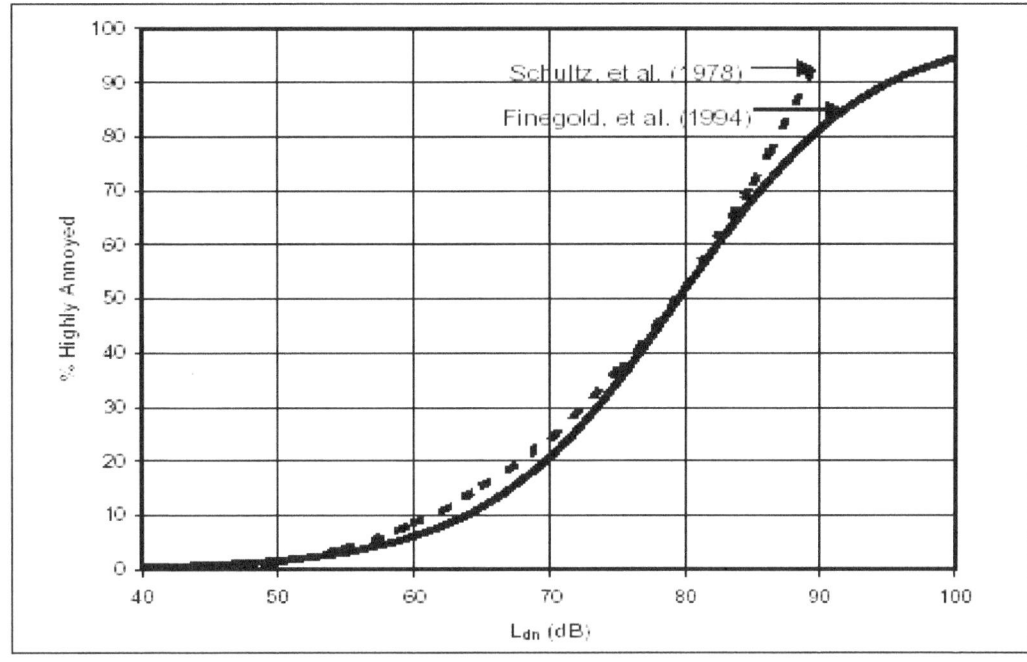

Source: Juneau EIS, Environmental Consequences Chapt. 4 – June 2001

Table 8: Hourly Traffic Volumes Based on Noise Levels

Max. Hourly Noise (dba)	Maximum Vehicles per Hour	
	Urban	Suburban
65 - daytime	320 vph	750 vph
60 - average	130 vph	260 vph
55 - nighttime	42 vph	74 vph

Source: ITE, Environmental Capacity of Neighborhoods 1996

4.6 Current and Future Considerations

There are several issues that are difficult to address in this analysis but need to be identified for consideration in future studies. These issues can be categorized by land use changes, airport use, changes in Rte 2A and neighboring roadways, current use and possible expansions of other modes like public transportation. Politics and funding are the last two key ingredients to consider identifying and implementing improvements in and around the Rte 2A corridor. In order to solve the current transportation problems, it is important to understand what the planning process is, who the stakeholders are and what role the airports and public transportation play in helping to solve the problems identified in this report.

4.6.1 Planning Process through MAPC

The Metropolitan Area Planning Council (MAPC) is the regional planning agency representing 101 cities and towns in the metropolitan Boston area. Created by an act of the Legislature in 1963, it serves as a forum for state and local officials to address issues of regional importance. The MAPC web site provided the details below on what planning process were going on and whom the players were.

MAPC works with its 101 cities and towns through eight subregional organizations. The subregions meet on a regular basis to discuss and work on issues of local concern. Concord, Acton, Bedford, Boxborough, Carlisle, Hudson, Lexington, Lincoln, Littleton, Maynard, and Stow are the communities of the Minuteman Advisory Group on Interlocal Coordination (MAGIC) subregion. Most community representatives are elected officials.

In 2001 MAGIC
- hosted two very successful, well-attended breakfasts with legislators from the region;
- participated in the development of the new Regional Transportation Plan, the main document that will determine transportation investments and funding until 2025, by reviewing the region's existing conditions, policies, and growth management options;
- launched the "MAGIC Carpet" study of alternative transportation opportunities in the subregion's eleven communities;
- discussed the implications of potential growth as shown by buildout analyses completed by MAPC in four MAGIC towns (Lincoln, Hudson, Acton, and Stow);
- hosted a workshop on Conservation Subdivision Design, a model study completed by MAPC on an innovative land use technique to preserve land while accommodating development; and
- reviewed Developments of Regional Impact, including Brookside Shops in Acton.

4.6.2 Planning Process through HATS

The Hanscom Area Towns Committee (HATS) closely monitors and evaluates the aviation related policy and associated development at Hanscom Field and the land and roads proximate to this aviation facility. The Committee is comprised of the Selectmen from Concord, Bedford, Lexington, and Lincoln, and also includes the active involvement and support of the Minuteman National Historic Park. Beginning in July, Concord served as the host community for HATS' meetings, chaired by Concord Selectman Gary Clayton.

In 2000, the Selectmen representing the four towns decided that the present level and scope of ongoing and proposed activities at Hanscom Field now mandate a higher level of coordination with the Hanscom Field Advisory Commission. In 2001, Selectmen representing the four towns continued to serve on both Hanscom-related committees.

Proposals by several private companies to expand the scope of existing commercial aviation activity at Hanscom Field were carefully scrutinized by the committee. The four towns continued to express their great concern with Massport's policy to promote Hanscom as a commercial aviation facility. In comments to members of the HATS Committee, the Massport Board of Directors maintained they cannot or will not limit the number of operations at Hanscom Field. Currently, within the framework of the 1995 Generic Environmental Impact Report, commercial airline activity is "limited" to 48 operations per day. Beyond that level, Massport must conduct an assessment of the environmental impact of additional commercial aviation operations. In related actions, the HATS Committee voted to terminate the litigation it had initiated against Massport in 1999. The four communities had contended that Massport's authorization to make Hanscom Field available to Shuttle America, Inc., a commercial aviation business, violated the provisions of M.G.L. c. 290, the legislation that established the Hanscom Field Advisory Commission. In 2000, Shuttle America also requested approval from the Federal Aviation Administration (FAA) to initiate service to La Guardia Airport in New York City. The FAA conducted a 30-day review period to solicit and evaluate additional information on the effect of the flights on various historic properties, including Minuteman National Historic Park. FAA's subsequent action to permit these new commercial flights was challenged by the Towns of Concord, Lexington, Lincoln and numerous private citizens in an administrative appeal brought to the U.S. Court of Appeals for the First Circuit. In 2001, the Court did not rule in favor of the plaintiffs, thus dismissing the towns' assertion that the FAA decision did not appropriately follow the review procedures as required by the National Historic Preservation Act.

The Noise Working Group met with the HATS Committee and proposed that the Group continue to exist due to ongoing studies. It was agreed that the Group be reconstituted with many of the same members, and that it remain under the joint supervision of HATS and Massport.

In 2001, Massport filed the draft scope of review for the Environmental Status and Planning Report (ESPR) for Hanscom Field with the State Executive Office of Environmental Affairs. This action formally initiates a detailed, comprehensive yearlong planning study and assessment of aviation related activities at Hanscom Field. The Town of Concord, through its Board of Selectmen, commented extensively on this draft public document. Massport has selected Rizzo Associates, Inc. as its prime consultant for this study that it conducts once every five years. The plan will evaluate past and current aviation related operations and their impacts. More critically, this plan will define the future scope of general and commercial aviation at Hanscom Field for the next ten years, or more.

Massport will again provide limited funding support to HATS, allowing it to retain its own consultants to assist in the towns' review of the ESPR. The HATS consultants will also work closely with and support the HATS environmental subcommittee, a group of volunteers with specialized knowledge in environmental site planning and analysis. While the tragic events of September 11 have generated significant turmoil and uncertainty with the aviation industry, Massport intends to move forward in 2002 with the ESPR process. There are still many questions and concerns the HATS Committee has regarding the credibility of any projections concerning the scope and nature of airline activity in the future.

The Hanscom Area Towns Committee (HATS) closely monitors and evaluates the aviation related policy and associated development at Hanscom Field and the land and roads proximate to this aviation facility. The Committee is comprised of the Selectmen and their representatives from Concord, Bedford, Lexington and Lincoln, and also includes the active involvement and support of the Minuteman National Historic Park. Lexington served as the host community for HATS' meetings this year, chaired by Lexington Selectman Peter Enrich.

Increased commercial as well as general aviation activity at Hanscom this year required HATS to be active in a variety of significant issues. In 1999, the Four Town Planning Group, which was established by HATS, developed a proposal for a process to review developments of regional impact (DRI). The purpose of creating a DRI review is to establish an intercommunity growth management process for the area where the four towns border each other. The DRI process will require each town to give notice of, review, and comment upon proposed significant projects, to develop and promote coordinated plans, programs and techniques of growth management; and, to coordinate efforts to evaluate proposed physical improvements to the region's infrastructure. The Planning Boards of the four towns adopted a Memorandum of Agreement codifying the DRI process in the fall of 2000.

The HATS Committee reviewed and adopted a comprehensive policy statement addressing the broad range of regional growth management issues associated with general and commercial aviation activity at Hanscom. In particular, the policy articulates that in the absence of a long-range regional smart growth strategy, there must be a moratorium on any additional commercial aviation, changes of use, and new infrastructure at Hanscom. This policy has also been adopted by several of our congressional representatives, the region's state senators and representatives, and the four towns' Boards of Selectmen.

The Selectmen representing the four towns decided that the present level and scope of ongoing and proposed activities at Hanscom now mandates a higher level of coordination with the Hanscom Field Advisory Commission (HFAC). During this past year, the four towns' Boards of Selectmen appointed a member of their respective Boards to HFAC to facilitate and enhance coordination among the varied interests at Hanscom.

In 1999, Massport voted to authorize making Hanscom Field available to Shuttle America, an airline flying fifty-seat airplanes. This vote was taken after little public review and substantial public opposition to this commercial activity. After exploring its legal options, and with support of the four Boards of Selectmen, HATS initiated a lawsuit against Massport. In July 2000, the towns requested a summary judgement by the State Superior Court based on the evidence before the courts that Massport had not complied with the provisions of M.G.L. c.290 - the legislation that established the Hanscom Field Advisory Commission. The judge did not rule in the towns' favor on this petition and the lawsuit remains pending for subsequent action by the Court.

This year also brought an expansion of commercial aviation activity at Hanscom when Shuttle America requested approval from the Federal Aviation Administration (FAA) to initiate service to LaGuardia Airport in New York City. The FAA conducted a 30-day review period to solicit and evaluate additional information on the effect of the flights on various historic properties, including Minuteman National Historic Park. FAA's subsequent action to permit these new commercial flights was challenged by the Towns of Concord, Lexington, Lincoln and numerous private citizens in an administrative appeal brought to the U.S. Court of Appeals for the First Circuit. The Court did not grant a request for a temporary injunction against the airline. A trial on the merits of this case remains pending. The towns continue to believe that the FAA decision did not appropriately follow the review procedures as required by the National Historic Preservation Act. At the same time, Massport continues to actively promote commercial aviation at Hanscom Field contributing to an ongoing, serious deterioration in the relationship between Massport and the affected communities.

Finally, initial discussions have commenced with Massport regarding the preparation of the next five-year Generic Environmental Report for Hanscom. This one to two year comprehensive planning study will evaluate past and current aviation related operations and their impacts. More importantly, it will define the future scope of general and commercial aviation at Hanscom Field in the years ahead.

4.6.3 Future Transportation Projects

The January 2001 Congestion Management System report inventoried congestion problem locations in MAGIC, as identified through field monitoring activities and comments from the public review process. Monitoring efforts identified long segments of Routes 2, 3, 4/225 in Bedford, Route 62, and I-95 as being congested, as well as shorter segments of Route 27 in Maynard, Route 85 in Hudson, Route 117 in Stow, Route 4/225 in Lexington, and the Lexington Minuteman Triangle area. Comments from responders resulted in the addition of long segments of Routes 2 and 2A in Lincoln, Concord and Acton, most of Route 27 through Acton, the Middlesex Turnpike, and numerous intersections to the list of identified congestion problem areas. This section from the Phase I Report for the Magic Study area addresses each location identified as congested and summarizes any recent activities or proposals for improvement. It also identifies locations, which may be considered for future study.

(1) Route 2 through Lincoln, Concord, and Acton
This section of the Route 2 corridor has experienced chronic congestion for years, principally because it remains a signalized arterial roadway, 2 lanes in each direction with turning lanes at key intersections. The remainder of Route 2 in both directions is a limited-access expressway facility. Several projects are currently underway to alleviate congestion in the

Route 2 corridor, including:

- grade-separation of the intersection of Route 2 with Cambridge Turnpike and Route 2A (Crosby's Corner) in Lincoln
- design studies to replace the existing rotary adjacent to the MCI facility in West Concord
- developer-sponsored coordination of traffic signals at 3 locations along Route 2 near the rotary
- additional safety improvements along the Route 2 corridor in Concord and Acton.

As mentioned earlier, analyses were already undertaken in past Route 2 corridor studies to evaluate the effects of grade-separation options at other signalized intersections. Such activities are probably the limit of what can be done to reduce congestion in the Route 2 corridor itself, because of agreements that stipulate that Route 2 cannot be widened beyond 2 lanes each way in this area. It may be possible to identify park-and-ride locations in the vicinity of the reconstructed rotary that would provide opportunities for carpool/vanpool formation and/or bus transit in this corridor.

(2) I-95, Lexington through the area of Route 3/3A exits
At present, no projects or proposals are active in this section of I-95 (Route 128) within Lexington, other than resurfacing. Alternatives, which might be examined in an attempt to address congestion in this section, could include:

- ramp metering at one or more interchange entrances
- investigation of park-and-ride lots in the area of the existing interchanges
- adding to existing Route 128 shuttle services operated by the Route 128 Transportation Task Force
- interchange improvements at the Route 4/225 and Route 3 interchanges

3) Route 4/225, Bedford and Lexington

The jughandle intersection with Hartwell Avenue, Lexington, continues to be a chronic congestion location along Route 4/225. In addition, the area near the intersection of Route 4/225 and Route 62 experiences long backups and delays as Route 62 becomes a more important conduit to the expanding Middlesex Turnpike industrial area. This location would probably operate more efficiently and more safely if the intersection was signalized. However, long delays will probably remain the norm here, if additional capacity is not provided. It may be that pressures on this intersection will be reduced as a result of Route 3 improvements, which should make it less desirable for Route 3 traffic to seek alternate routes through Bedford. In addition, the intersection of Route 4/225 at Shawsheen Road was identified as a congestion problem.

(4) Route 2A, Acton, Concord and Lexington
The entire corridor was identified as having congestion problems. In addition, the section of Route 2A/119 through Littleton is also reported to experience congestion problems. In Acton, Route 2A operates as one wide lane in each direction. Signalized intersections are infrequent, but large sections of the Route 2A corridor have been developed as commercial strip areas. Consequently, there are high concentrations of commercial driveways along the roadway, and vehicles making left turns into these driveways disrupt traffic flow. The town planner expressed interest in a corridor study of Route 2A aimed at solving the left-turn problem by adding left-turn lanes at selected locations, or creating a 3-lane cross- section including a center left-turning lane. Such a study should also include Route 2A/119
through Littleton.

In Concord, Route 2A shares the right-of-way with Route 2 for part of its length, and also shares in the problems of that roadway. The rotary adjacent to the Concord MCI facility creates delays for Route 2A traffic, because Route 2 traffic inside the rotary leaves few gaps for eastbound Route 2A traffic. This problem will be addressed as part of the overall design solution for the rotary and Route 2 in this area.

A limited reconstruction of Route 2A in the vicinity of Massachusetts Avenue and Forbes Road, and projects are currently included on the TIP list to improve bicycle connections in and around Hanscom Drive, the principal entrance to Hanscom Field from Route 2A. However, no other proposals or projects are currently under consideration for congestion reduction in this area.

In addition to these ongoing and prospective activities at locations identified as problem areas in the 2001 Congestion Management System report for MAGIC, other studies and initiatives are currently in planning or design. Projects currently listed in the regional Transportation Improvement Program (TIP), Table 9.

Table 9: TIP Projects Constructed or Advertised

1	Acton /Littleton Great Road resurfacing 601363
2	Acton/Boxborough/Littleton Route 2 improvements 601222
3	Bedford/Billerica/Burlington Route 62/Middlesex Turnpike widening 602079
4	Bolton Route 117 reconstruction 600580
5	Concord Pine St Bridge replacement 600638
6	Concord/Lexington/Lincoln Stone wall on Route 2A 601844
7	Hudson Chapin Road Bridge 160040

Table 9 continued: TIP Projects Programmed

8	Hudson Central St Hudson 600575
9	Lexington Marrett Rd Lexington, 043550
10	Lexington Lowell St-Woburn St 600883
11	Lexington Rt 2/I-95 (Route128) Interchange 601391
12	Lexington Depot Square improvements 602353
13	Lexington/Lincoln Route 2 median barriers 602308
14	Littleton Rt 119/Powers Rd Signal 600540
15	Maynard 3 intersections near town center 601577
16	Maynard/Stow White Pond Rd Bridge 601946
17	Bedford Depot Park improvements 602346
18	Bedford/Billerica/Burlington Crosby Drive widening/Rt 3 ramp 029490
19	Bedford/Billerica/Burlington Crosby Drive/Middlesex Turnpike 029491
20	Concord Monument St Bridge 601442
21	Concord Route 2 reconstruction 602626
22	Concord/Lincoln Crosby's Corner 602984
23	Concord/Lincoln Rt 117 Bridge 603079
24	Hudson Broad St Bridge 601907
25	Hudson/Marlborough Assabet Rail Trail 602947
26	Lincoln Battle Road Trail Underpass ---
27	Maynard & Acton Maynard Shuttle ---
28	Concord Main St Landscaping ---

Source: Magic Phase I Report

4.6.4 Public Transportation

(1) Commuter Rail

The Fitchburg commuter rail line is the principal transit route serving MAGIC commuters destined for downtown Boston and intermediate stations in Waltham and Cambridge. The Fitchburg Line has two tracks between Boston and South Acton; from South Acton westward, service operates on a single track. Four AM peak-period trips depart daily from Fitchburg beginning at 5:45 AM, with a fifth trip starting from South Acton at 8:41 AM. All trips stop at Waltham and Cambridge (Porter Square), but the 6:55 AM trip skips local stops in Weston, Waltham and Belmont.

The existing service is primarily oriented toward Boston core-bound commuters; but there is a great deal of interest among MAGIC communities in the potential for reverse commuting on the Fitchburg Line. Only limited service is available in the reverse-peak direction (2 outbound AM peak-period trips from Boston to South Acton, and 3 inbound PM peak period trips, also from South Acton). The recent CTPS reverse commute study, prepared for the MBTA, documents in great detail the constraints and costs associated with trying to reconfigure this line to provide additional service in the off-peak direction.

The main limitation of the existing service is the severe parking shortages that afflict all stations on the Fitchburg Line. There are only 1,075 spaces identified on the MBTA's web site for the entire line The largest parking area, at South Acton, is also located at the farthest origin station served by the last AM peak-period train. There are a total of 293 parking spaces at South Acton, but overflow parking can be seen on adjacent streets and unpaved areas.

There is resistance to the provision of additional commuter rail parking within several of the towns along the line, particularly Acton and Concord. However, the MAGIC towns are generally in support of the idea of a new regional station with large parking capacity, located on the outer edge of the MAGIC subregion. Residents of Littleton have proposed a new train station to replace the existing Littleton station; their preferred site is an environmentally sensitive parcel of land adjacent to Route 2 on the Littleton-Boxborough town line. Other towns, notably Acton and Concord, support this idea: for these towns, it represents a preferable alternative to parking expansions within their own borders. A new interchange connection to Route 2 would be required to serve a train station at this location.

(2) Bus Service

There are several different categories of bus service, with distinctly different markets in the Magic area. First, there are individual town-supported services. The Town of Lexington runs a mini-bus service (Lexpress), with the MBTA and the Town contributing subsidy support. Lexpress has a fairly dense route structure based in the Town center (Depot Square), and it runs hourly headways on all its routes from around 7:00 AM to around 6:00 PM (depending on route) daily. Also, Concord has its own town-supported local bus, operated with volunteer drivers. The bus runs a loop connecting several locations in Concord Center, also making one trip per day to one of the nearby supermarkets. The Town of Bedford operates a service that is essentially a demand-responsive van, which is booked like a taxi one day in advance. The van makes one trip per day to the Burlington Mall. Bedford was one of the first communities in which suburban transit services were tried, beginning in the late 1970s, then revived in 1985 after a hiatus. It, together with Lexpress and Burlington's B-line, continues to receive MBTA Suburban Transit funding to this day.

These services are open to anyone who wishes to travel, not just town residents. In fact, they tend to serve a market comprised primarily of transportation-disadvantaged town residents: people without cars available, elderly residents, people with physical disabilities which prevent them from driving, students needing transportation to after-school activities, and similar categories of people, all of whom represent good markets for transportation services. The services are generally less expensive to operate than traditional large-vehicle bus services, but they do require public subsidy to continue operating.

The second type of service is scheduled commuter service, operated by private carriers, the MBTA, and the Lowell Regional Transit Authority (LRTA). The Yankee Bus Company runs a single trip per day from Acton, via Concord, to downtown Boston and back; and the Gulbankian Bus Company, based in Southborough, runs service from Hudson to Boston via Marlborough and Framingham. These are the only direct private-carrier bus services in this area going to Boston.

However, the MBTA runs half-hourly service from Alewife Station to Lexington Center, Hanscom Field and the Bedford VA Hospital, and recently has added service to the growing Oak Park industrial area in Bedford. In addition, peak-period MBTA bus service to Government Center in Boston serves Burlington and Billerica, towns directly adjacent to MAGIC communities. Indeed, the bus stop within

the Sun MicroSystems campus in Burlington is the closest thing to a "regional transit node" in the MAGIC area, offering opportunities for passenger drop-offs and transfers among numerous services. Both LRTA and MBTA bus routes pick up and drop off at this location, and Burlington's mini-bus service (the B-Line) provides local circulation from this point.

(3) Shuttles

Paratransit-type shuttle services offered by the Route 128 Business Council and a handful of individual local companies offer opportunities for commuters in the MAGIC area. The Route 128 Business Council, a

Transportation Management Association (TMA) that serves employers in Waltham, Lexington and other locations along Route 128, is the largest operator of such services. The Council manages a comprehensive employer-supported shuttle service, using leased vehicles, which connects member employers with transit stations at Alewife, Riverside, Newton Highlands and several commuter rail stations. Their primary shuttle bus market consists of city-dwellers from Boston and Cambridge who take the Red Line to Alewife station, and then need a connection to their suburban work sites. Among the MAGIC-area companies, which are subscribers to these services, are:

- Fresenius Medical Care (Lexington)
- Hayden Woods (Lexington)
- Segue Software
- Spaulding & Slye
- Stride-Rite

As many as eight runs are made by these shuttles during the AM peak period, from Alewife out Route 2 to I-95 (Route 128). Fares are $2.00 per ride for employees of member companies, $4.00 per ride for others. Few individual employers offer such services for their own employees, because of the expense involved. Only a few companies/institutions within MAGIC operate their own vehicles, primarily for purposes other than employee commuting, but available to facilitate commuting as a side benefit, e.g., by providing access to remote parking sites, either within or outside their own property.

5.0 Optimal Traffic Volume

Using the results of the analysis on traffic volume, traffic congestion, traffic patterns, pedestrian access, and noise levels a traffic threshold of between 8,000 and 12,000 vehicle on an average day is proposed as a traffic volume for Rte 2A which would be "optimal" for a desired visitor experience and visitor safety.

The LOS at several intersections along Rte 2A is D or worse and will only get worse with time due to the anticipated growth in households, employment, and park visitation that are expected to occur during the next 20 years. If we assume that the minimal acceptable LOS is C, then between a 40 and 60% reduction in current traffic levels needs to occur to bring all of the intersections up to this level during the peak periods in the morning and in the evening without considering signalization.

The current v/c ratios at several points on the roadway during the peak hours of use are approaching their maximum value of 1.00. If we assume that a v/c is acceptable, then the traffic would need to be reduced by 40% to attain this level.

The current functional classification of Rte 2A favors mobility over access. If the park wants to promote access over mobility, a reclassifying of Rte 2A to a collector or a local with the necessary roadway improvements to ensure that it and the surround roads can handle the diverted traffic could reduce traffic by 50% to around 10,000 ADT. This is based on the relationship shown in Figure 18.

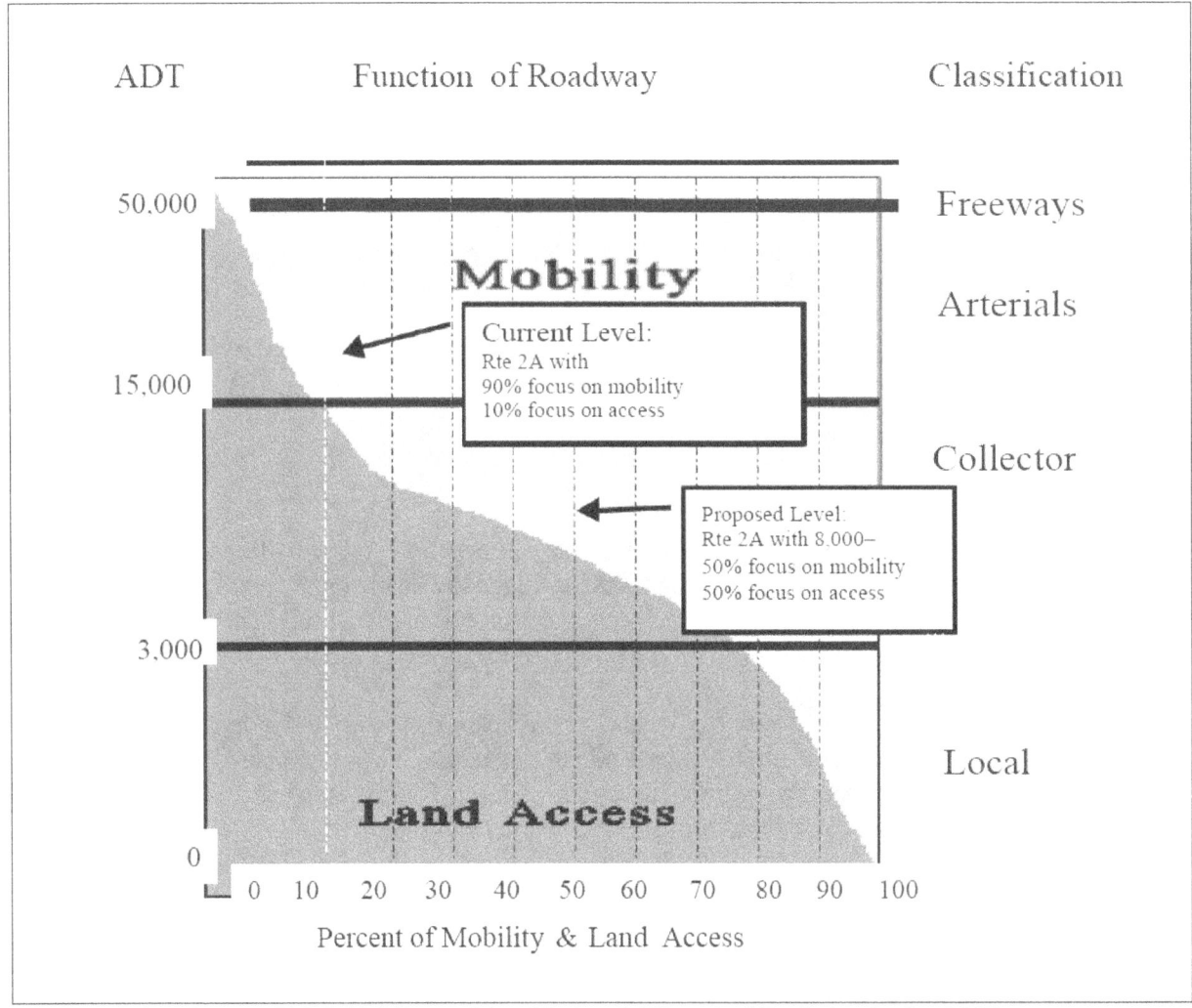

Figure 18: Mobility & Access Relationship

Source: Volpe Center (EG&G) Minute Man National Historical Park Analysis, 2002

The pedestrian delay shows that it would require several minutes to cross Rte 2A during the peak hours. If the park wanted to reduce this and promote activity on the southerly side of Rte 2A, a reduction of volume in access of 50% would be required to allow the park visitor to cross in a minute or two time span. The trip patterns have show that about 50% of the trips on Rte 2A are cut through trips, trips not originating or destined to anywhere in the study area. Noise was another factor in determining the optimal traffic volume range.

6.0 Alternatives

As Section 4.0 has articulated, Rt. 2A serves as a 'relief valve' to Rt. 2, with more than half of its ADT servicing through traffic beyond Crosby's corner to/from 95/128 and points east. It also services a substantial volume of traffic to/from the Air Force Base and commercial airport. Traffic to/from the Park is a minuscule portion of ADT; however, the impacts of Rt. 2A ADT are not inconsequential to the Park.

Options that would lessen these impacts – in particular, the severance and barrier effects of high-speed and high-volume traffic- must contend with a number of complicating factors that constrain feasibility and implementation. The Park needs to reach a negotiated agreement with other partners (most notably, Massachusetts Highway Department) since it neither owns nor manages the road. Financial constraints facing both the Park and its partners and lack of right-of-way favor operational and less capital-intensive options. Options that suppress traffic, other things being equal, are preferable to options that only divert traffic to other roads and neighboring towns. The ability of other roads in the network to absorb diverted traffic to the extent necessary with little and/or only minor design modifications is a key determinant of whether design and traffic management treatments for Rt. 2A are viable or not. Changes to the functional classification of the road (or at least change to its functional classification within the boundaries of the Park, seeking reclassification as a rural local access road) may be possible, but this too requires negotiated agreement with other partners (i.e., FHWA and MHD) and local businesses impacted by such changes. It is important to note, however, that the current range of ADT (i.e., 12-20,000 ADT) is consistent with its current functional classification as a rural minor arterial. It is possible to adhere to the current functional hierarchy for the regional road network (of which Rt. 2A is a part) with adoption of options that constrain traffic volumes to the lower threshold. This would substantially improve environmental quality of the corridor and lessen adverse impacts on the quality of the Visitor's experience.

Using the above noted factors as a screening and evaluative filter, we briefly describe several traffic management and design options that may be viable. We present these options separately, but as Section 4.1 outlines, the complexity of the problem dictates a need for a more holistic, integrated approach. This is sketched in Section 4.1 as a traffic management and road redesign plan in which the discrete elements work synergistically to achieve certain desired effects pertaining to the character of the road and corridor, and the environmental capacity of the road.

6.1 Rte 2A Redesign Options: Positive Impact on the Visitor Experience

Option 1 - This option[1] suggests a re-design of Rt. 2A (keeping in place the existing alignment) as one possibility for lessening the traffic impact on the Park and retaining more of the essential historic and visual character of the corridor.

The proposal is to alter the current cross-section of the road (see Figure 1(a)) to achieve a speed reducing 'traffic calming' effect, and concurrently to enhance the safety of the road. A useful byproduct of traffic calming tools that induce a speed-reducing effect is that somewhat modest volume reductions are also

---

[1] See D. Spiller, Conceptual Re-design of Rt. 2A, Technical Memorandum, 1/17/02 for further technical analysis and design rationale.

achieved. The modest volume reductions lessen the traffic impact of the road, but also are more easily absorbable by alternative routes, and less off-site facility expansion, if any, is needed.

Figure 1 (b), (c) and (d) show sketches of three variants of the same essential cross section redesign. The essential elements include: physically narrowing the travel lanes via incorporation of a continuous, centerline rumble strip^2 median and modest widening of the two shoulders, and additional 'optical' narrowing of the two travel lanes with a wide edge line marking. Each of these design elements is mutually reinforcing or synergistic in creating a holistic redesign of the cross section of the road.

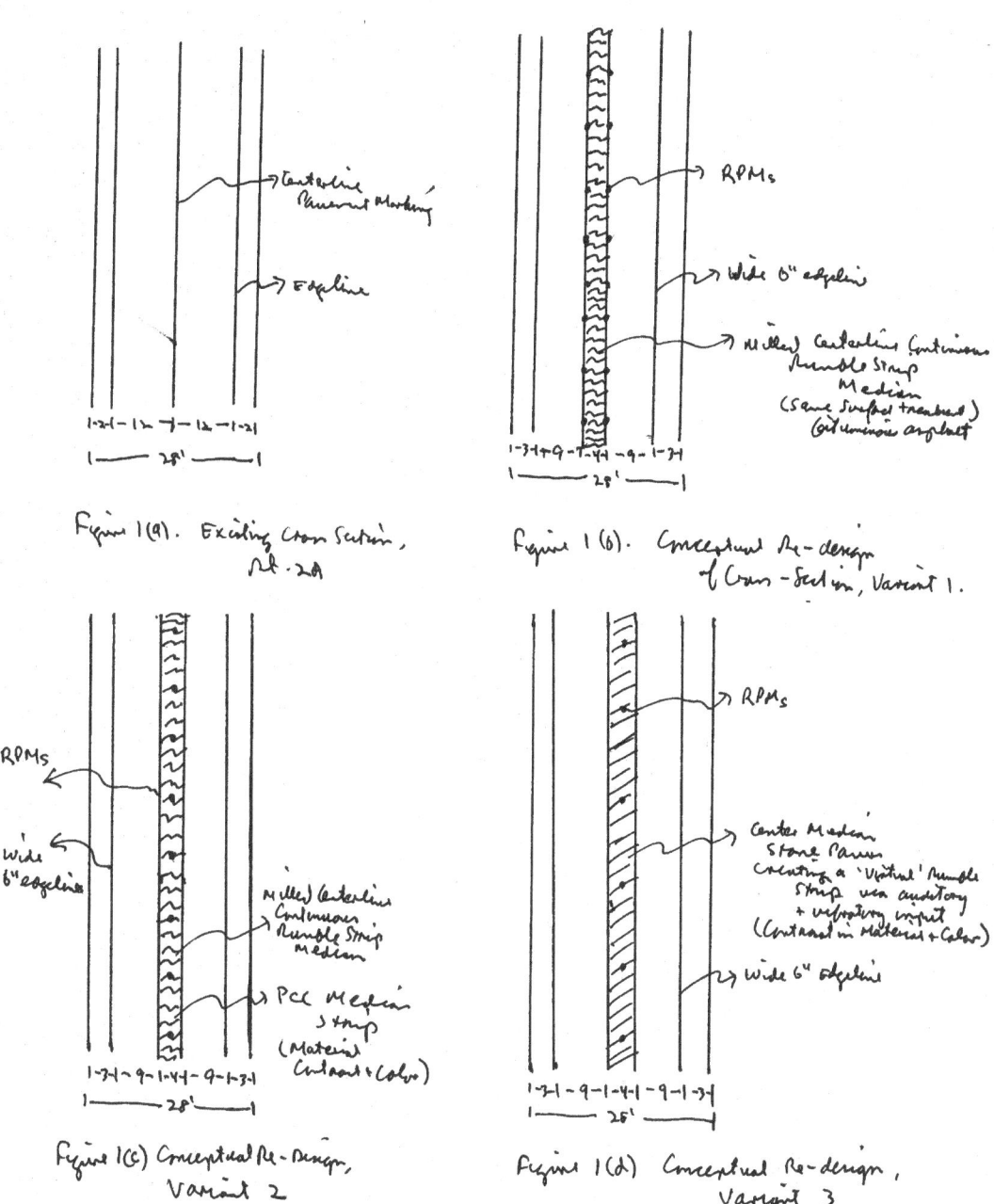

effect accentuates the speed difference between the two routes, inducing at least some volume of traffic to stay on and/or divert to Rt. 2.

Secondly, road safety is improved. The purpose of the continuous centerline rumble strip median as a design element is twofold. First, to reduce the accident risk of head-on collisions (one of the essential selling points to the Massachusetts Highway Department (MHD). Secondly, to provide a mechanism for physically narrowing the width of the travel lane and induce speed reducing, traffic calming effects from the side friction of the median (one of the essential selling points for the National Park Service). While it is useful to induce speed-reducing effects on the traffic stream, it is important to do so in a uniform way to keep speed variance low. This is in order to assure that the overall accident risk potential of the roadway configuration is tolerable and not increased over baseline conditions. Recent accident studies are starting to challenge conventional wisdom by demonstrating that narrower roadway and lane widths actually reduce injuries and fatalities, other factors held constant. An added advantage of the median treatment is that it provides a pedestrian refuge at major pedestrian crossings.

Thirdly, a major advantage is in the improvement to the 'soundscape' of the Park. Applications that place the rumble strips transverse to the road generate a noise event on each vehicular passage. The median treatment proposed here would generate noise events (affecting the 'soundscape' of the Park) on only rare occasions (i.e., upon vehicular encroachment of the median). Most importantly, reducing vehicular speeds, because of the nonlinearity of noise emissions with respect to speed, definitively improves the 'soundscape' of the Park.

Lastly, lowered speeds permit better and safer access to/from the gateways and parking lots to the Park.

Disadvantages – This option would require capital funds to reconstruct the cross section along an approximately 3 mile corridor. There would be temporary construction impacts (disruption, noise, added truck movements). Furthermore, agreement would need to be reached with MHD who owns and manages the road. The continuous centerline rumble strip may detract from the historic character of the corridor, particularly the abutting historic structures. This is partially mitigated, however, in that views from the Park and the historic structures across the road alignment are oblique. Furthermore, a variant of the design (see, e.g., Figure 1(d)) postulates only a 'virtual rumble strip' using stone pavers that would be contextual to the extant stonewalls along the trace alignment of the Battle Road.

Option 2 – This option proposes a partial or full closure (two sub options) of the Rt. 2A bypass, that portion of Rt. 2A between the junction of Rt. 2A and Rt. 2 (Crosby's Corner) and the junction of Rt. 2A with Lexington Road. The partial closure would be during the AM and PM peak hours.

Advantages – Probably no other option would be as effective in reducing the volume of traffic on Rt. 2A through the Park. More than fifty percent (53%) of two-way AM peak hour traffic on Rt. 2A originates from or is destined to points on Rt. 2 west of Crosby's corner. More than forty five percent (45.8%) of two-way PM peak hour traffic on Rt. 2A originates from or is destined to points on Rt. 2 west of Crosby's corner. No more than 1.4 percent of two-way ADT traffic on Rt. 2A (a potential vehicle two-way flow of 280 cars) is visitation traffic to the Park. Thus, the great majority of traffic on Rt. 2A is either other local access traffic within the corridor, and through traffic with points of origin and destination beyond the Park corridor.

Eastbound traffic on Rt. 2 with a desired interchange with 95/128 or destined to points further east would remain on Rt. 2. When Rt. 2 is congested, this traffic now diverts via the Rt. 2A bypass. The reverse traffic patterns, that is traffic coming off of the 95/128 interchange or from origins east of the interchange often use Rt. 2A as an alternative route to reach points west of Crosby's corner. From a purely mobility perspective, MHD has designed and implemented the 'perfect' connecting link in a regional road network to serve as a 'relief valve' to Rt. 2!

Disadvantages – The diversion of approximately 10,000 ADT (600-800 vehicles in the peak hour) would clearly be the greatest challenge. Could Rt. 2 handle the additional traffic load? Would additional capacity be needed? Would additional capacity be needed along an extended length of the route, or only at specific bottleneck locations? Should no improvements be made, what would be the deterioration in level-of-service (LOS)? What other traveler behavioral responses would be likely (e.g., suppression of trips, shifting to other routes, changes in departure times, peak spreading, shifts to other modes, etc.)? What would be the order of magnitude (i.e., volume of trips affected) for these behavioral responses?

Should only a partial closure be implemented, what would be the operational costs for 'gating' operations? How would advance communications to motorists for a time-dependent closure be handled? What signage and ITS systems would be necessary?

Option 3a & 3b - These options suggest an improvement to the junction of Rt. 2A, Lexington Road and Brooks Road by incorporating either a mini-roundabout[3] as a traffic control device or realigning Lexington Road so that it forms a T intersection. This spot location has experienced a relatively high rate of accidents, in part due to the high approach speeds and poor sight lines resulting from the skewed angle approaches. The rationale for proposing this option includes: enhanced safety of traffic at the junction; achieving speed-reducing, traffic calming effects; providing a clearer definition of priority of way, and preservation of historic character by avoiding traffic signalization. As Section 5.1 elaborates, this option would be quite complementary to the median treatment under Option 1.

Option 3a - At this point, this option is only a proposal that would require further planning study to determine whether construction of a roundabout is feasible, that it can service the traffic efficiently and safely, and fits within the park's mission to preserve and enhance the historical character of the road. Site-specific factors, including right-of-way limitations, utility conflicts, drainage problems, adverse grades, etc., may preclude its suitability.

Advantages – With proper design principles – flared (i.e., wider) approaches, deflection of traffic at entry, and yield on entry to circulating traffic – the intersection would operate safer while still handling the traffic demand. Sight lines would be improved. Approach speeds would be substantially reduced. Traffic flow would be uninterrupted, and delays would be lower than with traffic signals at the same traffic volumes handled by the junction.

---

[3] Mini-roundabouts are distinguished from traditional roundabouts primarily by their smaller size and compact geometry. They are typically designed for negotiation speeds of 25 km/h (15 mph). Inscribed circle diameters generally vary from 13 m to 25 m (45 ft. to 80 ft.) See, Roundabouts: An Informational Guide, FHWA-RD-00-067, June 2000.

The mini roundabout could also serve as a landscape treatment providing a gateway to the Park for visitors traveling eastbound from the North Bridge and Merriam's Corner. It would signal a change in character of the roadway, complementary to that achieved by the median treatment under Option 1.

Disadvantages – This option would involve capital costs. Disruptions would occur during construction. While the total amount of queuing of traffic on the approaches may be no more (and likely to be less) than for alternative traffic control treatments such as signalization, the queues of traffic on the Rt. 2A approaches may be longer because the mini roundabout operates on a FIFO regimen. With a traffic signal, additional 'green' time for the priority approaches of Rt. 2A would reduce the queues on the approaches.

Option 3b - This option is only a proposal that would require further planning study to determine whether construction of a T Intersection is feasible, and that it can service the traffic efficiently and safely. Site-specific factors, including right-of-way limitations, utility conflicts, drainage problems, adverse grades, etc., may preclude its suitability.

Advantages – the intersection would operate safer while still handling the traffic demand. Sight lines would be improved. Approach speeds would be substantially reduced. Traffic flow on Rte and delays would be lower than with traffic signals at the same traffic volumes handled by the junction.

Disadvantages – This option would involve capital costs. Disruptions would occur during construction. While the total amount of queuing of traffic on the approaches may be no more (and likely to be less) than for alternative traffic control treatments such as signalization, the queues of traffic on the Rt. 2A approaches may be longer on the Lexington Road segment.

Option 4 – This option proposes to impose turning restrictions at the two junctions with Rt. 2A that have level-of-service (LOS) D or F during peak hours. These junctions are the intersection of Bedford Road with Rt. 2A, and the intersection of Hanscom Drive with Rt. 2A. It is proposed under this option that left turns be prohibited on the eastbound approach at Hanscom Drive, and on the westbound approach with Bedford Road.

Advantages – By restricting left turns from the eastbound approach of Rt. 2A onto Hanscom Drive, motorists seeking access to the Air Base and commercial airport will be diverted to alternative routes outside the boundary of the Park (e.g., Virginia Road and Hartwell Drive). Similarly, restriction of the left-turning movement onto Bedford Road from the westbound approach of Rt. 2A prevents the use of Rt. 2A as a 'cut-through', 'rat-running'[4] alternative to Rt. 2 for traffic originating from the 95/128 interchange and heading westbound to points west of Crosby's corner.

Safety will be improved. First, the risk of rear-end collisions will be reduced since high-speed traffic will not have to decelerate quickly upon joining the back-of-the-queue of traffic on the eastbound approach blocked by a left-turning vehicle. Second, left-turning vehicles will not have to accept less than adequate gaps in the opposing traffic stream precisely for the fear of getting rear-ended.

---

[4] British term!

Queuing of traffic, with adverse localized air emission concentrations, will be reduced on the Rt. 2A approaches to both junctions. This will also reduce delays to Park visitors caught in the queues seeking access to the Park downstream from both junctions.

Disadvantages – Since one person's gain is often another person's loss, imposing traffic management regulations via turn restrictions modifies the traffic pattern and reduces easy access to the airport or to Rt. 2 via Rt. 2A.

Unlike self-enforcing traffic calming designs, enforceability of the turn restrictions may be problematic. This could cause even greater safety issues since the expectancy is changed with respect to the uniformity of motorist behavior.

Option 5: Identifying locations for Crosswalks to improve visitor access

6.2 Designation Status Options as a Means to Improve Rte 2A

The state of Massachusetts and the Federal Highway Administration both allow for roads to have special designations as a means of promoting their use, protecting their character, and providing special funding to repair and enhance their appearance. Communities, tribal governments, State and Federal agencies, and other groups can organize information on special roads in their areas and nominate a road for national status. To do so requires the following:

- Completing an inventory
- Making a strong case or evaluation to determine why a road belongs in the national program
- Assembling materials such as maps, photos, interpretation and marketing ideas
- Preparing a corridor management plan that meets national criteria[5]
- Getting an endorsement from the state scenic byway program
- Seeking designation from the Secretary of Transportation

Option 6 - Massachusetts has no specific administrative process or technical criteria for designating a road as a scenic byway[6]. Rather, a local group and/or coalition of communities through which the road passes would make such designation via passage of an Act of the Legislature as a result of a direct lobbying effort. Clearly, the positions taken by MHD and MASSPORT in support of or in opposition to such a designation would be critical to the success of this effort.

It is almost certain the case that Route 2A, within the limits of the Battle Road Unit of the Park, could meet the two tests for intrinsic historic quality: there are enough sites and locations within the corridor to create a story with a certain level of continuity and coherence. The historic story also provides a link among resources at various points along the byway and a means of interpreting these resources to the Visitor. The essential historic content is the start of the Revolutionary War with the movement and counter-movement of opposing forces in a series of engagements along the general alignment of the

---

[5] National Scenic Byways Program, Federal Register, Vol. 60, No. 96, Thursday May 18, 1995, pp. 26759-26762.
[6] Personal communications, Sarah Bradbury, MHD, Bureau of Transportation Planning and Development; there is a Scenic Roads program under Chapter 40, Section 15C that does provide 'historic preservation' protection against road modifications that would impinge on roadside trees and stone walls. It applies, however, to only local roads, that is, non State-designated routes.

road. The physical elements of the landscape within the Battle Road unit corridor, which comprise its historic features, can be inventoried, mapped and interpreted. The elements would comply with FHWA interim policy with respect to possession of integrity of location, design, setting, material, workmanship, feeling and association. With the establishment now of the Battle Road Trail within the corridor and adjacent to the road, it is also possible that recreational qualities and even archeological qualities (there are several active dig sites within the Park) are significant and unique enough to qualify as well.

Option 7 - For national Scenic Byway designation, the road and its corridor need to possess at least one of six intrinsic qualities: archeological, cultural, historic, natural, recreational, and scenic. The elements that comprise a corridor management plan per FHWA policy for designation under the National Program are summarized in Table 1. It is important to note that the National Scenic Byways program has a dual purpose of protection and promotion. Communities have sought designation not only because of the community-enhancing aspects of pride and sense of place often engendered from the planning process of seeking national recognition, but also from specific tangible benefits that potentially derive from designation[7]:

- Increased business, tax revenue and jobs from tourist dollars
- Federal and state funding for planning and following through on a corridor management plan
- Increased regional cooperation among stakeholders
- Protection for a resource that the community believes is threatened, including a higher sensitivity among local residents to the value and potentially fragile nature of the landscape along the byway
- A better understanding of the byway corridor's ability to absorb change
- A more even approach to addressing needs of various roadway users
- Improved maintenance and a higher budget for the road. For example, some states give scenic byways a higher priority for shoulder work and road repair than other highways, and the maintenance they get is of higher grade.
- Access to resources and expert assistance in managing the corridor.
- Identification on state, federal and auto club highway maps, leading to more tourism opportunities for the area.
- Money and other assistance from state and national offices of economic development and tourism.

The process of establishing a corridor management plan in the context of seeking national scenic designation imposes somewhat narrow limits on the extent of the corridor to be considered and subject to preservation actions. The corridor in this context extends only as far as the viewshed from the road. This would generally be a narrow swath, particularly given the relatively dense foliage along some segments, and would most definitely exclude for example Hanscom AFB and the civil airport. Yet, the problems that the Park faces include land uses and potential development outside the Park's boundaries, not to mention outside the byway's corridor. Similarly, the current and proposed arterial and highway network as well as potential regional transit services are also factors determinative of what the future impact on the Park is. Noise, visual intrusion, severance effects, and access delay at the Park's entrance

---

[7] See US DOT, FHWA, Community Guide to Planning and Managing a Scenic Byway, 2000; and FHWA and National Park Service's Rivers, Trails and Conservation Assistance Program, Byway Beginnings: Understanding, Inventorying, and Evaluating a Byway's Intrinsic Qualities, 1999. Both these documents provide excellent guidance and checklists for engaging in a participatory and collaborative planning process for pursuing national scenic designation.

gates and parking facilities from the traffic generated from intensification of development in the region, including access to/from the base and civil airport, is what is most problematic about Route 2A from the perspective of the Park. The Park's interests in seeking to restore and preserve the intrinsic historic character of the byway and corridor extend beyond an emphasis on structures to include the historic setting and association. This argues for strategies and management actions, including alternative road design modifications[8], which reduce the flow of traffic on the byway.

It is quite possible, and even advantageous, for Minuteman NHP to engage in a dialogue with potential partners, stakeholders, and the public as part of a collaborative and participatory planning effort. The organizing framework could be a corridor management plan, but one not limited to the boundaries and strategies implicit in the criteria for the national scenic byway program. A good example is the US Route 7 Corridor Transport Plan[9] completed under the auspices of the Chittenden County Metropolitan Planning Organization (Burlington, VT). The Park could then assure that the focus remains as:

- Increased regional cooperation among stakeholders
- Protection for a resource that the community believes is threatened, including a higher sensitivity among local residents and state agencies to the value and potentially fragile nature of the landscape along the byway
- A better understanding of the byway corridor's ability to absorb change and accommodate traffic, with an accompanying management action plan to assure that the traffic flow does not exceed the environmental capacity of the road or impair its historic character
- A more even approach to addressing needs of various roadway users

Table 1. FHWA Required Components for Corridor Management Plan under the National Scenic Byway Program

1. A map identifying the corridor boundaries, location of intrinsic qualities, and land uses in the corridor.
2. An assessment of the intrinsic qualities and their "context" (the areas surrounding them).
3. A strategy for maintaining and enhancing each of these intrinsic qualities.
4. The agencies, groups, and individuals who are part of the team that will carry out the plan, including a list of their specific, individual responsibilities. Also, a schedule of when and how you'll review the degree to which those responsibilities are met.
5. A strategy of how existing development might be enhanced and new development accommodated to preserve the intrinsic qualities of your byway.
6. A plan for on-going public participation.
7. A general review of the road's safety record to locate hazards and poor design, and identify possible corrections.
8. A plan to accommodate commercial traffic while ensuring the safety of sightseers in smaller vehicles, as well as bicyclists, joggers, and pedestrians.
9. A listing and discussion of efforts to minimize anomalous intrusions on the visitors' experience of the byway.

---

[8] See, e.g., D. Spiller, Conceptual Re-design of Route 2A, Technical Memorandum, 1/17/02.
[9] See http://www.ccmpo.org/planning/rt7/index.htm.

10. Documentation of compliance with all existing local, state, and federal laws about the control of outdoor advertising.
11. A plan to make sure that the number and placement of highway signs will not get in the way of the scenery, but still be sufficient to help tourists find their way. This includes, where appropriate, signs for international tourists who may not speak English fluently.
12. Plans of how the byway will be marketed and publicized.
13. Any proposals for modifying the roadway, including an evaluation about design standards and how proposed changes may affect the byway's intrinsic qualities.
14. A description of what you plan to do to explain and interpret your byway's significant resources to visitors.

6.3 Rte 2A Redesign Options With a Possible Negative Impact on the Visitor Experience

The Volpe Study Team was asked by Minute Man National Historic Park to briefly examine other potential alternatives to reduce traffic congestion and speed which would be useful from a traffic engineering perspective yet also possibly detrimental from a NPS resource protection perspective. Options 8 through 11 may at some point be suggested by the Mass Highway Department or Massport as potential alternatives to manage traffic but are seen to be in conflict with preserving the historical nature of Route 2A.

Presently Route 2A operates approximately 85% capacity. As traffic flow increases the LOS declines. This decline is worsened by traffic turning off 2A and traffic entering 2A. Traffic making left turns off 2A is especially detrimental. Two types of factors may cause a decline in LOS. They are traffic volume and traffic movements. Increases in traffic volume decrease the LOS when the number of lanes, roadway geometry and conditions are inadequate to allow the required number of vehicles to flow at safe and desired speeds. Traffic movements such as traffic entering and exiting the roadway (right turns), and left turns cause the otherwise steady flow to become unsteady. The effect of this second condition increases as the volume to capacity (V/C) ration increase.

Assuming that land use and traffic patterns stay the same, as the traffic volumes increase the level of service will decrease. Mitigation measures to maintain and acceptable LOS may include adding additional lanes, adding queuing lanes for left turns, deceleration lanes for traffic exiting Route 2A, accelerations lanes for traffic entering 2A.

The following is a discussion of the effects, pros and cons of each of these measures.

Option 8 - Adding an additional lane or lanes in each direction mitigates the first factor (traffic volume) affecting LOS. Realignment (geometric modifications) of route 2A may also be proposed. These modifications would likely be minor because roadway geometry does not appear to be at issue on 2A except at Lexington Road.

Additional lanes would allow an increase in traffic volume without lowering the LOS.

Advantages

- Allow Route 2A to handle a higher volume of traffic with an acceptable LOS
- Allow speed variations of traffic flowing in the same direction.
- Allow passing without entering into oncoming traffic. Although passing may be allowed on multi lane roadways where conditions permit.

Disadvantages

- Increase in impervious area that will require storm water management.
- Increased traffic volumes will cause an increase in noise. This is especially noticeable during off peak times when the lower volumes allow higher speeds.
- Pedestrians wanting to cross Route 2A will have to cross-additional lanes.

Option 9 – Construct queuing lanes for left turns (with or without islands) at Bedford Street and Hanscom Drive. These lanes are auxiliary lanes that address the second factor (traffic movements) affecting LOS. They work by maintaining steady flow in the through lane or lanes. Queuing lanes

Advantages

- Traffic turning left from route 2A will not interfere with through traffic provided that the LOS of the oncoming traffic remains high.
- Less additional impervious area than with adding and entire additional lanes.
- Less intrusive to the surrounding infrastructure. If a median were part of the plan, it would provide pedestrian refuge.

Disadvantages

- Increase in impervious area.
- Does not address making left turn against oncoming traffic unless traffic signals area installed.
- Queuing lane locations will likely coincide with pedestrian crosswalks to minimize flow interruption to traffic. However this also means that pedestrian have a longer distance to cross.
- Roadway pavement area must flare to accommodate the auxiliary lanes. This may cause unsteady flow near the intersection. If a median is part of the plan, then flaring may not be required. Instead the median will neck down to accommodate the additional lane. From a right of way and pavement area standpoint, the median would be the equivalent of adding an additional lane.

Option 10 - Acceleration/Deceleration Lanes also act as auxiliary lanes that address the second factor (traffic movements) affecting LOS. They work by maintaining steady flow in the through lane or lanes.

Advantages
- Deceleration lanes allow right turning traffic to leave the main traffic stream before slowing down to turn. Design of these lanes can provide some queuing for traffic turning right.
- Acceleration lanes allow traffic entering the main traffic stream to attain the running speed of the main traffic stream. This also provides the entering traffic greater flexibility to merge.

Disadvantages
- Increase in impervious area
- Increased distance for pedestrians to walk. This alterative would not likely require islands. Therefore if pedestrian refuge were provided, it would require addition land.

Option 11 - Traffic signals are typically installed based on warrants such as accident data. They address the second factor affecting LOS. However, if improperly designed, they can decrease the LOS for through traffic due to excessive interruptions.

Advantages

- Provide for orderly traffic movement
- Sometimes they can increase the traffic capacity of the intersection
- Reduce accidents, especially front to side collisions
- Interrupt heavy traffic to allow traffic from side streets to cross or turn and pedestrians to cross.

Disadvantages (especially if improperly designed)

- Cause excessive delays
- If signals are improperly designed, disobedience may increase thereby increasing the possibility of accidents
- Rear end collisions can increase
- Use of alternate routes incapable of handing additional traffic may be encouraged to avoid the traffic lights.
- Masts, arms and control boxes will be visually detrimental to the character of Route 2A and the park.

General Mitigation Plan

Any plan to mitigate traffic impacts from commuters, the airport or alternative development on the airport property will likely be a combination of some or all of the above measures. For example, an addition lane in each direction could also function as an acceleration/deceleration lane. Further analysis is required to properly select and evaluate actual mitigation measures.

6.4 Framework for a Traffic Management and Road Re-design Plan

Although Section 6.1-6.3 outlines alternative options, the problem of dealing with Rt. 2A in the context of the Park and corridor and the regional road network is more complex than a choice between discrete options. The data and analysis in Section 4 suggest a framework for a traffic management and road re-design plan that integrates several of the articulated options in a hybrid, holistic strategy that is likely to yield greater speed-reducing and volume-reduction effects than selection of any one option singularly. We sketch such an articulation of a Plan below.

There appears to be too much network connectivity and permeability, allowing Rt. 2A to serve as an alternative connecting route for through traffic from the west to an interchange with 128/95 and points east of the circumferential highway, as well as the reverse movement. Closure of the Rt. 2A bypass (the segment of Rt. 2A from Rt. 2 at Crosby's corner to the junction of Rt. 2A with Lexington Road) during peak hours would sever the existing network connectivity and associated permeability of traffic and substantially reduce the traffic loads on Rt. 2A. Permanent closure would achieve very substantial volume reduction in traffic using Rt. 2A but probably at too high a congestion cost to the regional network.

Redesign of the cross-section of Rt. 2A to incorporate a continuous centerline median (whether an actual continuous centerline rumble strip, or a 'virtual rumble strip' using stone pavers) would be complementary to the Rt. 2A bypass closure. By narrowing the traffic lanes, it would bring the design speed in line with a more desirable operating speed. This would reduce head-on accident risk, improve access to the Park, and facilitate more convenient and safer pedestrian crossings. It also would signal a change in character of the road relative to its competing route (i.e., Rt. 2). It would move the functional transformation of Rt. 2A toward a local access road, serving Park access as well as other corridor development. Further study of this option is required to understand the design specifications required to preserve the historical character of the road

Complementary to a centerline median redesign could be a roundabout treatment at the junction of Rt. 2A, Rt. 2A-bypass road, Brooks Road and Lexington road to solve a particularly unsafe spot location that has experienced a high accident rate. With proper deflection upon entry, improved sight lines and with the approach speeds lowered by the narrower travel lanes, the junction would operate more efficiently and, most importantly, safer. Again further study is required to understand the design requirements to preserve the cultural resources and views.

Of all the options proposed in this report the Volpe Study Team recommends, as a first step, for the National Park Service to work with its partners and local stakeholders to form a working group that would develop a corridor management plan as described in section 6.2. This planning could be done in conjunction with obtaining Scenic Byway designation at the state and federal level or done separately. Because of the volume of traffic on route 2A, the regional importance of is road as connector as well as the connection to Hanscom Airport and Air Force base it is critical for all the stakeholders to work together on the future use of the road. A corridor management plan is the proper forum for finding such agreement.

REPORT DOCUMENTATION PAGE		Form Approved OMB No. 0704-0188

The public reporting burden for this collection of information is estimated to average 1 hour per response, including the time for reviewing instructions, searching existing data sources, gathering and maintaining the data needed, and completing and reviewing the collection of information. Send comments regarding this burden estimate or any other aspect of this collection of information, including suggestions for reducing the burden, to Department of Defense, Washington Headquarters Services, Directorate for Information Operations and Reports (0704-0188), 1215 Jefferson Davis Highway, Suite 1204, Arlington, VA 22202-4302. Respondents should be aware that notwithstanding any other provision of law, no person shall be subject to any penalty for failing to comply with a collection of information if it does not display a currently valid OMB control number.
PLEASE DO NOT RETURN YOUR FORM TO THE ABOVE ADDRESS.

1. REPORT DATE (DD-MM-YYYY) 05/2002	2. REPORT TYPE Planning Study	3. DATES COVERED (From - To) NA
4. TITLE AND SUBTITLE Minuteman National Park: Rte 2a Traffic Analysis and Its Impact on the Park's Visitor Experience		5a. CONTRACT NUMBER NA
		5b. GRANT NUMBER NA
		5c. PROGRAM ELEMENT NUMBER NA
6. AUTHOR(S) Jeff Bryan, David Spiller, Scott Peterson (EG&G), Francis Ford (EG&G)		5d. PROJECT NUMBER
		5e. TASK NUMBER NPS TIC No. D-97
		5f. WORK UNIT NUMBER NA
7. PERFORMING ORGANIZATION NAME(S) AND ADDRESS(ES) U.S. Department of Transportation Research and Special Programs Administration John A. Volpe National Transportation Systems Center		8. PERFORMING ORGANIZATION REPORT NUMBER DOT-VNTSC-NPS-05-43
9. SPONSORING/MONITORING AGENCY NAME(S) AND ADDRESS(ES) National Park Service Alternative Transportation Program 1201 Eye St. NW Washington, DC 20005		10. SPONSOR/MONITOR'S ACRONYM(S) NERO
		11. SPONSOR/MONITOR'S REPORT NUMBER(S) (see 5d. and 5e. above)

12. DISTRIBUTION/AVAILABILITY STATEMENT
Public distribution/availability.

13. SUPPLEMENTARY NOTES
This report addresses alternative transportation decision factors as indicated below (Y/N/NA):
(N) Non-construction options; (Y) park carrying capacity; (N) life-cycle/ops. & maintenance costs; (N) cost-effectiveness.

14. ABSTRACT
This work supports Minuteman National Park (the Park) development of an amendment to the 1991 General Management Plan by identifying the past present and possible future levels of traffic occurring on the Battle Road (Rte 2a) and then determining what is an acceptable level of traffic along this historic roadway. To help lessen the impact of traffic, several ways of minimizing traffic along this roadway will be investigated in order to produce the most realistic and beneficial solution to the Park and the visitor experience.

15. SUBJECT TERMS
Minuteman National Park, Battle Road, Lexington, Concord, Lincoln, Rte 2a, Hanscom Airfield

16. SECURITY CLASSIFICATION OF:			17. LIMITATION OF ABSTRACT	18. NUMBER OF PAGES	19a. NAME OF RESPONSIBLE PERSON Gary T. Ritter
a. REPORT	b. ABSTRACT	c. THIS PAGE	NA	61	19b. TELEPHONE NUMBER (Include area code) 617-494-2716, ritter@volpe.dot.gov
None	None	None			

Standard Form 298 (Rev. 8/98)
Prescribed by ANSI Std. Z39.18